IFIP Advances in Information and Communication Technology 418

IFIP – The International Federation for Information Processing

IFIP was founded in 1960 under the auspices of UNESCO, following the First World Computer Congress held in Paris the previous year. An umbrella organization for societies working in information processing, IFIP's aim is two-fold: to support information processing within its member countries and to encourage technology transfer to developing nations. As its mission statement clearly states,

> IFIP's mission is to be the leading, truly international, apolitical organization which encourages and assists in the development, exploitation and application of information technology for the benefit of all people.

IFIP is a non-profitmaking organization, run almost solely by 2500 volunteers. It operates through a number of technical committees, which organize events and publications. IFIP's events range from an international congress to local seminars, but the most important are:

- The IFIP World Computer Congress, held every second year;
- Open conferences;
- Working conferences.

The flagship event is the IFIP World Computer Congress, at which both invited and contributed papers are presented. Contributed papers are rigorously refereed and the rejection rate is high.

As with the Congress, participation in the open conferences is open to all and papers may be invited or submitted. Again, submitted papers are stringently refereed.

The working conferences are structured differently. They are usually run by a working group and attendance is small and by invitation only. Their purpose is to create an atmosphere conducive to innovation and development. Refereeing is also rigorous and papers are subjected to extensive group discussion.

Publications arising from IFIP events vary. The papers presented at the IFIP World Computer Congress and at open conferences are published as conference proceedings, while the results of the working conferences are often published as collections of selected and edited papers.

Any national society whose primary activity is about information processing may apply to become a full member of IFIP, although full membership is restricted to one society per country. Full members are entitled to vote at the annual General Assembly, National societies preferring a less committed involvement may apply for associate or corresponding membership. Associate members enjoy the same benefits as full members, but without voting rights. Corresponding members are not represented in IFIP bodies. Affiliated membership is open to non-national societies, and individual and honorary membership schemes are also offered.

Andreas Burg   Ayşe Coşkun
Matthew Guthaus   Srinivas Katkoori
Ricardo Reis (Eds.)

VLSI-SoC:
From Algorithms to Circuits
and System-on-Chip Design

20th IFIP WG 10.5/IEEE International Conference
on Very Large Scale Integration, VLSI-SoC 2012
Santa Cruz, CA, USA, October 7-10, 2012
Revised Selected Papers

 Springer

Volume Editors

Andreas Burg
EPFL, Lausanne, Switzerland
E-mail: andreas.burg@epfl.ch

Ayṣe Coṣkun
Boston University, MA, USA
E-mail: acoskun@bu.edu

Matthew Guthaus
University of California, Santa Cruz, CA, USA
E-mail: mrg@soe.ucsc.edu

Srinivas Katkoori
University of South Florida, Tampa, FL, USA
E-mail: katkoori@cse.usf.edu

Ricardo Reis
Universidade Federal do Rio Grande do Sul
Porto Alegre, Brazil
E-mail: reis@inf.ufrgs.br

ISSN 1868-4238   e-ISSN 1868-422X
ISBN 978-3-642-45072-3   e-ISBN 978-3-642-45073-0
DOI 10.1007/978-3-642-45073-0
Springer Heidelberg New York Dordrecht London

Library of Congress Control Number: 2013953903

CR Subject Classification (1998): C.5.4, B.7, C.3, C.1, C.0, B.8, B.6, B.7

Typesetting: Camera-ready by author, data conversion by Scientific Publishing Services, Chennai, India

Printed on acid-free paper

Springer is part of Springer Science+Business Media (www.springer.com)

Preface

This book contains extended and revised versions of the best papers that were presented during the 20th edition of the IFIP/IEEE WG10.5 International Conference on Very Large Scale Integration, a global System-on-a-Chip Design & CAD conference. The 20th conference was held at the Dream Inn Hotel, Santa Cruz, California, USA (October 7–10, 2012). Previous conferences have taken place in Edinburgh, Trondheim, Vancouver, Munich, Grenoble, Tokyo, Gramado, Lisbon, Montpellier, Darmstadt, Perth, Nice, Atlanta, Rhodes, Florianópolis, Madrid, and Hong Kong.

The purpose of this conference sponsored by IFIP TC 10 Working Group 10.5, the IEEE Council on Electronic Design Automation (CEDA), and by IEEE Circuits and Systems Society, with the In-Cooperation of ACM SIGDA, is to provide a forum for the exchange of ideas and presentation of industrial and academic research results in the field of microelectronics design. The current trend toward increasing chip integration and technology process advancements has brought about stimulating new challenges both at the physical and system design levels, as well as in the test of these systems. VLSI-SOC conferences aim to address these exciting new issues.

The 2012 edition of VLSI-SoC maintained the traditional structure of the conference, which has been successful at the previous VLSI-SOC conferences. The quality of submissions (110 regular papers and nine special session papers from 15 countries) made the selection process difficult. Finally 33 papers were accepted for oral presentation and 17 posters were accepted for presentation. Out of the 33 regular oral papers presented at the conference, 12 papers were chosen by a selection committee to have an extended and revised version included in this book. The selection of these papers has considered the evaluation scores during the review process and the review forms provided by members of the Technical Program Committee and session chairs as a result of the presentation. The chapters of this book have authors from Belgium, Brazil, China, Italy, Sweden, Switzerland and the USA. The Technical Program Committee comprised 97 members.

VLSI-SoC 2012 was the culmination of the work of many dedicated volunteers: paper authors, reviewers, session chairs, invited speakers and various committee chairs, especially the local arrangements organizers. We thank them all for their contribution.

This book is intended for the VLSI community, mainly those persons who did not have the chance to attend the conference. We hope you will enjoy reading this book and that you will find it useful in your professional life and for the development of the VLSI community as a whole.

October 2013

Andreas Burg
Ayşe Coşkun
Matthew Guthaus
Srinivas Katkoori
Ricardo Reis

Organization

The IFIP/IEEE International Conference on Very Large Scale Integration-System-on-Chip (VLSI-SoC) 2012 took place during October 7–10, 2012, in the Dream In Hotel, Santa Cruz, California, USA. VLSI-SoC 2012 was the 20th in a series of international conferences, sponsored by IFIP TC 10 Working Group 10.5 (VLSI), IEEE CEDA, and ACM SIGDA.

General Chair

Matthew Guthaus   UC Santa Cruz, USA

Program Chairs

Ayse Coskun   Boston University, USA
Andreas Burg   EPFL, Switzerland

Special Sessions Chair

Wentai Liu   UC Santa Cruz, USA

Local Arrangements Chair

Jose Renau   UC Santa Cruz, USA

Publication Chairs

Srinivas Katkoori   Univ of South Florida, USA
Ricardo Reis   UFRGS, Brazil

Publicity Chair

Ricardo Reis   UFRGS, Brazil

Registration Chair

Rajsaktish Sankaranarayanan   UC Santa Cruz, USA

Finance Chair

Baris Taskin   Drexel, USA

PhD Forum Chair

Ken Pedrotti   UC Santa Cruz, USA

Web Chair

Walter Condley   UC Santa Cruz, USA

Steering Committee

Chi-Ying Tsui   HKUST, Hong Kong, SAR China
Manfred Glesner   TU Darmstadt, Germany
Luis Miguel Silveira   INESC ID, Portugal
Salvador Mir   TIMA, France
Ricardo Reis   UFRGS, Brazil
Michel Robert   University of Montpellier, France

Table of Contents

FPGA-Based High-Speed
Authenticated Encryption System

Michael Muehlberghuber, Christoph Keller,
Frank K. Gürkaynak, and Norbert Felber

Integrated Systems Laboratory (IIS), ETH Zurich,
Gloriastrasse 35, 8092 Zurich, Switzerland
{mbgh,chrikell,kgf,felber}@iis.ee.ethz.ch

Abstract. The Advanced Encryption Standard (AES) running in the
Galois/Counter Mode of Operation represents a de facto standard in
the field of hardware-accelerated, block-cipher-based high-speed authenticated encryption (AE) systems. We propose hardware architectures
supporting the Ethernet standard IEEE 802.3ba utilizing different cryptographic primitives suitable for AE applications. Our main design goal
was to achieve high throughput on FPGA platforms. Compared to previous works aiming at data rates beyond 100 Gbit/s, our design makes
use of an alternative block cipher and an alternative mode of operation,
namely Serpent and the offset codebook mode of operation, respectively.
Using four cipher cores for the encryption part of the AE architecture,
we achieve a throughput of 141 Gbit/s on an Altera Stratix IV FPGA.
The design requires 39 kALMs and runs at a maximum clock frequency
of 275 MHz. This represents, to the best of our knowledge, the fastest
full implementation of an AE scheme on FPGAs to date. In order to
make the design applicable in a real-world environment, we developed a
custom-designed printed circuit board for the Stratix IV FPGA, suitable
to process data with up to 100 Gbit/s.

Keywords: Authenticated encryption, High-throughput architecture,
FPGA, Pipelining, Serpent, OCB, AES, GCM.

1   Introduction

Confidentiality and authenticity are two of the most important cryptographic
goals. Whereas the former assures that any eavesdropping adversary is unable to
decipher a given message—even if she has access to the transmission medium—,
the latter refers to the cryptographic service that ensures that the receiver of a
message can be sure about its origin, i.e., that an attacker has not impersonated
the sender. Authenticated encryption (AE) combines these two services and
allows a *secure* and *authentic* communication between two parties.

In order to provide high-throughput AE implementations based on block ciphers, so-called *combined* modes of operation have been designed throughout
the last decade. They allow a higher throughput by interleaving the authentication part and the encryption part instead of calculating them consecutively

A. Burg et al. (Eds.): VLSI-SoC 2012, IFIP AICT 418, pp. 1–20, 2013.

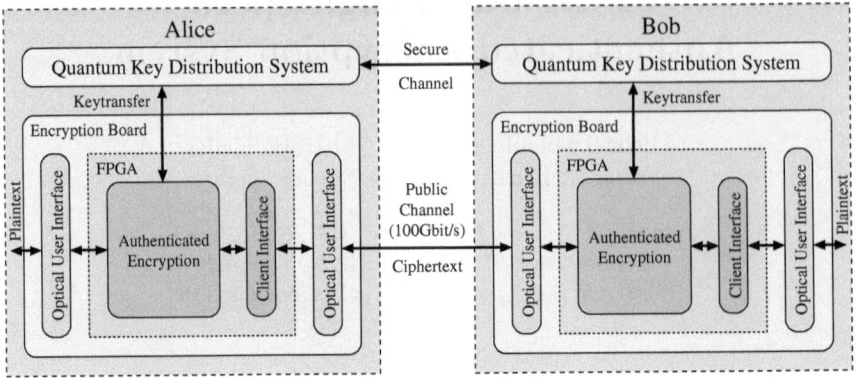

Fig. 1. High-speed authenticated encryption system setup

(as traditional AE methods do). The two most widely accepted AE modes of operation are *Counter with CBC-MAC (CCM)* [19] and *Galois/Counter Mode (GCM)* [11]. Their acceptance is most likely due to the fact that they have been recommended by the National Institute of Standards and Technology (NIST) (cf. [5] and [6]). Since then, they have been applied to technologies and protocols such as WiFi 802.11 [8] and IPsec [17]. Although the specifications of these modes do not determine the underlying block cipher, most applications make use of the Advanced Encryption Standard (AES) [14] since it is another algorithm standardized by the NIST.

The present work proposes a block cipher-based hardware architecture for AE, targeting high throughput on field-programmable gate array (FPGA) platforms. Our design has been developed as to fulfill the requirements of the Ethernet standard IEEE 802.3ba [1], which allows for transmission speeds of up to 100 Gbit/s. This work has been designed as part of a system that employs quantum key distribution (QKD) for synchronizing multiple private key exchanges within a single second, and provides authenticated encryption service using conventional cryptographic primitives. Fig. 1 illustrates the overall system setup. The main contributions of our work are related to the *Authenticated Encryption* part of Fig. 1, i.e., the digital, AE-related parts on the FPGA and have originally been presented in [12].

So far, our system employed a common GCM-AES-based cryptographic primitive in order to achieve the required throughput. In this work, we examine alternatives for both the block cipher and the mode of operation and compare the performance of these alternatives to the established cryptographic primitives. Besides exploring more efficient hardware implementations, this work is also motivated by providing an alternative AE scheme, in case successful attacks are developed against the existing primitives. We evaluate the Serpent block cipher [3] and the offset codebook (OCB) mode of operation [16] and we provide results of hardware implementations for different mode of operation/block cipher combinations, namely:

- OCB-Serpent   - GCM-Serpent
- OCB-AES   - GCM-AES

Our fastest AE implementation is based on an OCB-Serpent architecture and requires 39 kALMs (Adaptive Logic Modules) on an Altera Stratix IV FPGA. It uses four cipher cores for the encryption part and reaches a throughput of 141 Gbit/s, running at 275 MHz.

Moreover, we developed a custom-designed printed circuit board (PCB), which allows us to use the presented designs in real-world applications such as the system illustrated in Fig. 1. So far, two copies of the board have been fabricated and successfully tested in various sample experiments.

The remainder of this work is structured as follows. In the next section, we present an overview of related work on hardware architectures targeting high-throughput AE designs. In Section 3, a description of Serpent and OCB is given. The actual hardware architecture of our design is presented in Section 4. Throughout Section 5, we summarize our results, including a brief discussion. Finally, Section 6 provides a description of the custom-designed PCB including some of its major features, before we conclude our work in Section 7.

2   Related Work

Due to the standardization by the NIST, GCM-AES has received significant attention from both the research community and the industry, and several implementations targeting FPGAs can already be found in the literature.

In 2009, Zhou et al. [20] presented a single-core GCM-AES design, which targets a Xilinx Virtex-5 FPGA. They achieved a throughput of 41.5 Gbit/s using the 128-bit version of AES. Henzen and Fichtner [7] showed that it is possible to break the 100 Gbit/s barrier on a Virtex-5 FPGA platform. They made use of four fully unrolled AES cores for the encryption part and used four Karatsuba-Ofman (KO) multipliers in order to realize the authentication part. Their design reaches a throughput of 119.3 Gbit/s.

The most complex operation during the computation of a message digest according to GCM is the multiplication in the binary finite-field $GF(2^{128})$, which is part of the universal hashing function called GHASH. Therefore, most of the effort in improving GCM implementations has been spent on speeding up this calculation. Wang et al. [18] presented a GHASH architecture based on four GHASH cores that achieved a throughput of 123.1 Gbit/s on a Virtex-5. Crenne et al. [4] reached 238.1 Gbit/s by using 8 parallel finite-field multipliers, also targeting a Xilinx Virtex-5 FPGA. Since we aim at a full AE architecture, i.e., a design including both the authenticity and the confidentiality part, we do not consider these GHASH-only implementations for our investigations.

To the best of our knowledge, no hardware architecture based on a block cipher other than AES and targeting a high-throughput AE implementation has been presented so far. Moreover, no AES design has been published to date, which makes use of an operation mode different than GCM in order to achieve throughputs up to 100 Gbit/s.

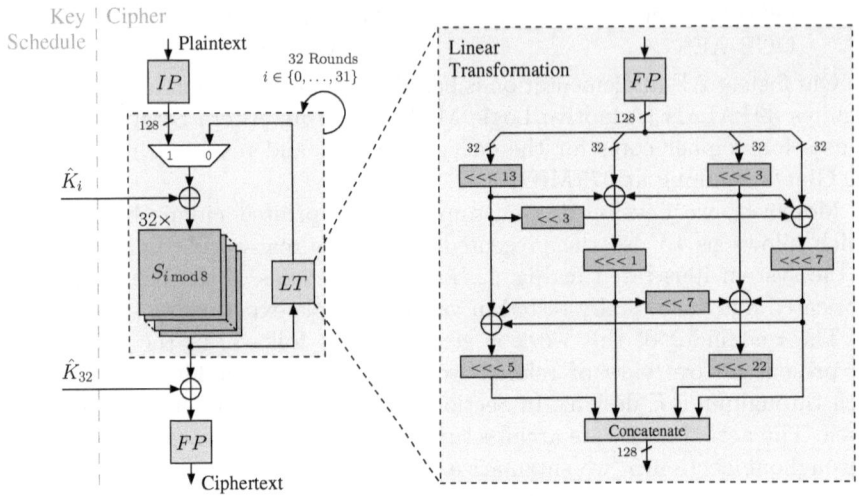

Fig. 2. Serpent block cipher

3   100 Gbit/s Authenticated Encryption Alternatives

In order to reach throughputs exceeding 100 Gbit/s on commercial FPGA devices, it is necessary to use multiple parallel instances of cryptographic primitives. Although AES running in GCM mode is currently the most widespread option for high-throughput hardware architectures, using these cryptographic primitives is not a requirement. Different block ciphers and modes of operation, like the ones presented in the following sections, can be used for a throughput-oriented AE system as well.

3.1   Serpent Block Cipher

Serpent was the runner-up of the AES block cipher competition. Although it has not been chosen by the NIST during the competition, it was considered to be a close alternative and is still known to be secure from a cryptographic point of view as the considerably large number of rounds contributes to its security [13]. In the following we will briefly discuss the main components of Serpent, i.e., the key schedule and the cipher itself, using the *conventional* implementation described in the official proposal [2]. In order to change from the *conventional* to the *bitslice* version of Serpent[1], all instances of the initial and the final permutation have to be omitted.

Cipher. Fig. 2 illustrates the Serpent cipher which consists of an initial permutation (IP), 32 round transformations, and a final permutation (FP). The

---

[1] We refer to the Serpent proposal [2] for further information on the *bitslice* implementation.

first 31 rounds of the cipher include a *key-mixing stage*, a *substitution stage*, and an *avalanche stage* (i.e., a stage where a linear transformation takes place). In the last round of the cipher, the linear transformation is omitted and replaced by another key-mixing operation. Serpent makes use of eight different S-boxes $(S_i, i \in \{0 \dots 7\})$ which repeat themselves every eighth round as shown in Fig. 2. Note that only a single S-box is used within each round of the cipher.

Key Schedule. The key schedule of Serpent takes a 256-bit cipher key K and expands it to thirty-three 128-bit subkeys denoted by $\hat{K}_i$. Cipher keys shorter than 256 bits are padded by appending a single "1", followed by as many "0"s as required in order to reach a length of 256 bits. After padding, K gets expressed using eight 32-bit values, i.e., $K = \{w_{-8}, \dots, w_{-1}\}$, and extended to an intermediate key $\{w_0, \dots, w_{131}\}$ according to

$$w_i = (w_{i-8} \oplus w_{i-5} \oplus w_{i-3} \oplus w_{i-1} \oplus \phi \oplus i) <\!\!<\!\!< 11\,, \qquad i \in \{0 \dots 131\}\,,$$

where $<\!\!<\!\!< i$ denotes a rotate-left function by i bits and $\phi = 0x9E3779B9$, i.e., the 32-bit value of the fractional part of the golden ratio. The actual subkeys, which are required during the round transformations of the cipher, are finally obtained by

$$\hat{K}_i = IP(S_{(3-i) \bmod 8}(w_{4i}, w_{4i+1}, w_{4i+2}, w_{4i+3}))\,, \qquad i \in \{0 \dots 32\}\,,$$

where S_i refers to one of the eight Serpent S-boxes. Similar to the cipher, the *bitslice* implementation of the Serpent key schedule can be obtained by removing all instances of the initial permutation IP. For a detailed description of Serpent, including additional information about the initial and the final permutation, we refer the reader to the official Serpent proposal [2].

3.2 Offset CodeBook Mode

The offset codebook (OCB) block cipher mode of operation is a combined AE scheme and has first been published by Rogaway et al. [16] in 2001. It is strongly related to the Integrity Aware Parallelizable Mode (IAPM) by C. Jutla [9] and three different versions have been made public since 2001. Throughout the remainder of this work we solely refer to the third version of it, i.e., OCB3 [10].

To start the authenticated encryption scheme according to OCB, a plaintext message, denoted by M, gets split into m different blocks, each of length n and an optional block M_* of length smaller than n as follows[2]:

$$M = \begin{cases} M_1, \dots, M_m\,, & \text{if } |M| = k \cdot n \text{ and } k \in \mathbb{N}, \\ M_1, \dots, M_m, M_*\,, & \text{else}\,. \end{cases}$$

Algorithm 1 and Fig. 3 describe the authenticated encryption according to OCB using pseudo-code and a block diagram, respectively. For simplicity, only

---

[2] We refer to the length of x in bits using the following notation: $|x|$.

Algorithm 1. OCB authenticated encryption

Input: Message M, Message block length n, Cipher key K, Nonce N, Associated data
  A, Authentication tag length τ, $0 \leq \tau \leq 128$
Output: Ciphertext C, Authentication tag T
1: **if** $|N| \geq n$ **then return** INVALID
2: $\{M_1, \ldots, M_m, M_*\} \leftarrow M$, with $|M_i| = n$ and $|M_*| < n$
3: $Checksum \leftarrow 0^{128}; C \leftarrow 0^{128}$
4: $L_*, L_\$, L[0] \ldots L[\lfloor log_2(m) \rfloor] \leftarrow Setup(K, m)$
5: $\Delta \leftarrow Init(N, n, K)$
6: **for** $i = 1$ to m **do**
7:     $\Delta \leftarrow \Delta \oplus L[ntz(i)]$ $\triangleright Inc_i(\Delta)$
8:     $C \leftarrow C \| E_K(M_i \oplus \Delta) \oplus \Delta$
9:     $Checksum \leftarrow Checksum \oplus M_i$
10: **end for**
11: **if** $M_* \neq \varnothing$ **then**
12:     $\Delta \leftarrow \Delta \oplus L_*$ $\triangleright Inc_*(\Delta)$
13:     $Pad \leftarrow E_K(\Delta)$
14:     $C \leftarrow C \| M_* \oplus (Pad \wedge (1^{|M_*|}))$
15:     $Checksum \leftarrow Checksum \oplus M_* 10^*$, with
            $M_* 10^* = M_* \| 1 \| 0 \ldots 0$, such that $|M_* 10^*| = n$
16: **end if**
17: $\Delta \leftarrow \Delta \oplus L_\$$ $\triangleright Inc_\$(\Delta)$
18: $Final \leftarrow E_K(Checksum \oplus \Delta)$
19: $Auth \leftarrow Hash_K(A)$
20: $Tag \leftarrow Final \oplus Auth$
21: $T \leftarrow trunc(Tag, \tau)$
22: **return** $C \| T$

---

the cases for full message blocks, i.e., $M_* = \varnothing$ is shown in Fig. 3. The cipher starts with a setup and initialization step (cf. line 4 and 5 in Algorithm 1). Thereafter, each message block can be processed independently of each other (line 6 to 16). Finally, the authentication tag T is determined throughout line 17 to 21. The characters $\|$, $\oplus$, and $\wedge$ represent the concatenation, bitwise exclusive or, and bitwise and operation. The term $ntz(i)$ describes the number of trailing zeroes of i in binary representation. 0^n and 1^n stand for bit strings of length n containing only zeros and ones, respectively. Furthermore, $trunc(X, y)$ truncates the bit string X to its y least significant bits. We use $\varnothing$ to represent an *empty set*. Appendix A provides listings for the *Setup*, *Init*, and $Hash_K$ procedures used throughout Algorithm 1.

When using a counter for the nonce N, the calculation of the initial offset Δ requires a block cipher call only every $64th$ initialization. This is due to the fact that the least significant six bits of N are set to zero before passing it to the block cipher (cf. line 3 of Algorithm 3). This fact together with the parallelizable processing of the message blocks, makes OCB suitable for high-throughput applications.

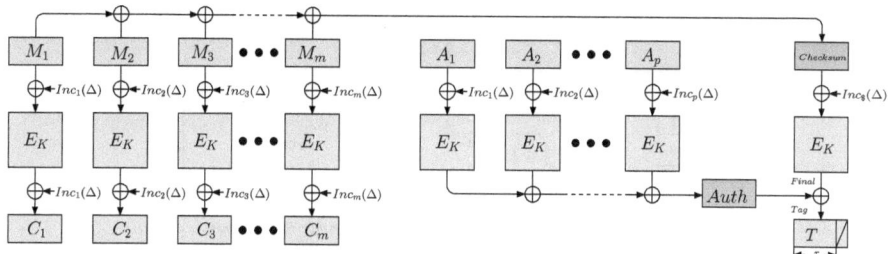

Fig. 3. Overview of the encryption and the authentication part of OCB

4   OCB-Serpent Hardware Architecture

For the OCB-Serpent architecture, supporting the IEEE 802.3ba Ethernet standard, we assume the following prerequisites:

- The size of the message block counter i is restricted to 7 bits, since 2^7 message blocks are capable of hosting a full Ethernet frame.
- As our target application ensures solely full message blocks, we do not handle short final message blocks separately.

As previously mentioned, to achieve extremely high throughputs, a multi-core approach has proven to be the only viable option when implementing AE on commercial FPGA platforms. Similar to GCM, OCB also allows two successive message blocks to be processed independently of each other. We have taken advantage of this fact and decided to use four parallel cipher cores in order to achieve the desired throughput. Fig. 4 illustrates the OCB architecture for authenticated encryption based on four Serpent cores.

4.1   Pipelined Four-Core Serpent Architecture

Each of the four Serpent cores handles a single 128-bit message block. Therefore, the overall design can process a 512-bit message at a time. As can be seen from Fig. 4, we fully unrolled the 32 rounds of Serpent. Furthermore, we inserted pipeline stages after each round in order to increase the maximum frequency of the cipher cores. Although the pipelined architecture allows us to clock the Serpent cores at a higher frequency, one problem inherent to all pipeline architectures has to be taken into consideration: When the normal flow of operations has to be suspended, the entire pipeline must be stopped in order to allow the rest of the operation to resume. Such an occurrence is known as a *pipeline stall* and can, for instance, occur during a key change.

Due to the unrolling of the Serpent rounds, we have to realize 1024 of the 4-bit S-boxes for each core, which requires a considerable amount of resources. The subkeys for the key-mixing stage of all four Serpent cores are provided from a single key schedule, i.e., the cores always operate on the same cipher key.

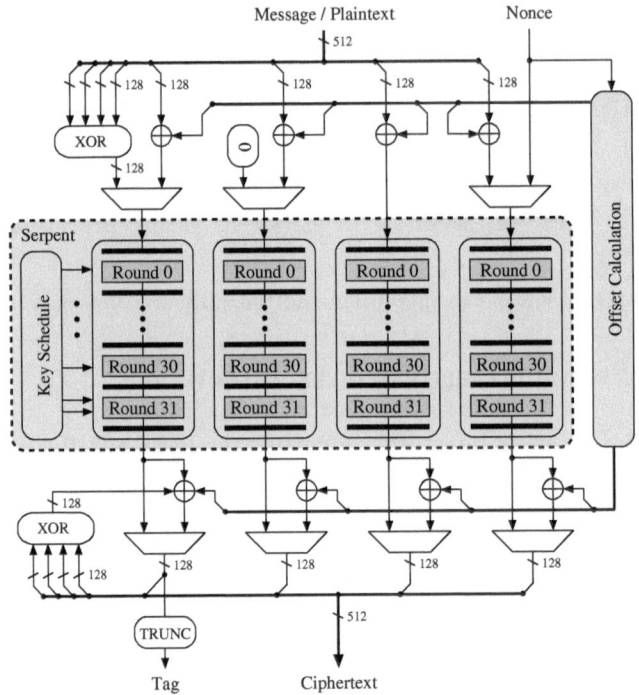

Fig. 4. OCB-Serpent architecture

4.2   OCB - Authenticated Encryption

OCB can, in general, be subdivided into three stages: *Initialization, encryption/authentication,* and *finalization*. During the initialization phase, two potential pipeline stalls may occur if not handled properly. First, after each key change a cipher call $E_K(0^{128})$ is required in order to compute the table values $L[..]$ (cf. Algorithm 2). Second, each new message needs a fresh nonce N, and thus a new offset value Δ. Since the calculation of the initial offset also requires a cipher call, this may result in another pipeline stall. In order to reduce the number of pipeline stalls to a minimum, we precompute the initial offset values.

The limitation of message lengths to a maximum of 2^7 blocks facilitates the precomputation of the $L[..]$-values, as it limits the maximum number of trailing zeroes of the block counter i to six. Thus, only $L_\$$, L_*, and $L_0 \ldots L_6$ have to be precomputed and stored in registers. In fact, when the result of $E_K(0^{128})$ is available, all nine table values can be computed in a single clock cycle[3].

When processing a block in authenticated-encryption mode, the message gets XOR-ed with the current offset Δ_i, encrypted, and finally XOR-ed with Δ_i again. As pipeline stages were introduced into the cipher cores, the Δ-values either have

---

[3] The calculation only depends on operations cheap to implement in hardware, i.e., fixed shift and conditional exclusive or operations.

to be stored or recalculated. We decided to recompute the offsets as it makes the implementation less dependent on the underlying block cipher and the number of pipeline stages used. Since the multi-core design processes four message blocks in parallel, the offset-calculation needs to be capable of providing four offset values at a time. In fact, the calculation of the four offset values is relatively cheap as the initial offset only has to be XOR-ed with the precomputed table values.

As described in Algorithm 3, a nonce-dependent call to the encryption of the block cipher is required. The result of this operation, further-on called $Ktop$, then has to be shifted by a 6-bit nonce-dependent value $Bottom$. First, in order to be able to perform this shift-operation, $Bottom$ has to be buffered until the result of $E_K(Top)$ is available. Second, the 6-bit variable shift is done using a 192-bit by 6-bit barrel shifter. Although using a counter for the nonce N could avoid the resource-expensive barrel shifter, we decided to keep it in order to stay independent of the actual structure of the chosen nonce.

4.3 Decryption

Authenticated decryption according to OCB is very similar to the encryption process. Exchanging the encryption operation E_K of the underlying cipher by the decryption operation D_K in line 8 of Algorithm 1, turns OCB into decryption mode. However, the other encryption operations remain. Therefore, the multi-core decryption unit contains four block-cipher decryption cores, one encryption core, and a common key schedule. In order to assure authenticity of a provided message, the re-calculated message tag T' must be equal to the tag T, received from the opposite communicating party.

A minor drawback of authenticated decryption according to OCB is the fact that a delay, dependent on the number of pipeline stages p between the calculation of the plaintext and the calculation of the authentication tag exists. This delay is caused by the calculation of the authentication tag which requires the $Checksum$ of the plaintext. Therefore, in order to verify the authentication tag of a message, p plaintext blocks have to be buffered, resulting in additional memory requirements.

5 Results

We coded our architectures in VHDL. For the synthesis and place&route design steps, we used Altera Quartus II version 11.0. Functional correctness was verified with Modelsim 6.6e simulator. Synthesis was conducted using a speed-optimzed setting. Except of M9K block memories, no Altera-specific logic blocks had been used. In order to have some reference implementations regarding AES and GCM on our target platform (Altera Stratix IV), we also synthesized a four-core AES cipher architecture as well as GCM based on both Serpent and AES. The AES architectures were accomplished with an underlying four-core AES similar to the one proposed by Henzen et al. [7]. We used the 128-bit version of AES and fully unrolled the 10 rounds. Furthermore, pipelining registers have been inserted after each round similar to the Serpent cipher core design.

Table 1. Encryption-only and authenticated encryption results targeting an Altera Stratix IV (EP4S100G5F45) platform using four cipher cores

Block Cipher	Mode of Operation	Area [ALMs]	[M9K Bl.]	f_{max} [MHz]	Throughput [Gbit/s]	[%]
Cipher-Only Architectures						
Serpent	cipher-only	28,399	0	281	144	136
AES	cipher-only	7,661	314	267	137	130
Authenticated Encryption Architectures						
Serpent	OCB	38,312	0	275	141	133
AES	OCB	11,948	314	250	128	121
Serpent	GCM	56,474	0	203	104	99
AES	GCM	24,313	314	206	105	100

The first two rows of Table 1 contain the place&route results of the multi-core encryption architectures (i.e., without running any mode of operation). The subsequent rows present the results for the different combinations of block ciphers and modes of operation. Regarding the block ciphers, the fully unrolled four-core AES design requires less area as it only has to provide 160 8-bit S-boxes for each cipher core, compared to the 1024 4-bit S-boxes needed by the Serpent cores. For the AES cores we utilized the M9K memory blocks in order to implement the 160 S-boxes, whereas for Serpent we implemented them solely in look-up tables. One of the reasons for this design decision was the significant routing overhead, which would have been required for the 1024 Serpent S-boxes being realized in M9K memory blocks. Since the high-throughput universal hash function GHASH of the GCM mode occupies a lot of area, OCB designs result in a smaller footprint than their GCM counterparts.

Regarding the throughput, all architectures met the target of 100 Gbit/s. However, the OCB versions are considerably faster. This is mainly because of the simpler architecture of OCB compared to GCM, which requires the resource-expensive GHASH function. Table 1 shows that the critical path of the GCM-based architectures is dominated by the authentication part, whereas the OCB-based designs almost reach the maximum frequency of the cipher-only implementations (with minor exceptions which are most likely due to placement and routing disparities). Compared to our results in [12], we were able to further increase the maximum frequency of both our cipher-only architectures as well as the designs based on OCB. Although our results show that OCB is at least as suitable for high-throughput hardware implementations as GCM, the latter is still the preferred mode of operation in the literature when designing high-speed authenticated encryption hardware architectures based on block ciphers. This might be due to the facts that there are some US patents on OCB and that, in contrast to GCM, it has not been recommended by the NIST. Regarding the patents on OCB, its author has recently eased licensing for a variety of

Table 2. Block cipher modes of operation comparison

Property	OCB	GCM
Patented	Yes	No
Parallelizable	Yes (Encr. + Auth.)	Yes (Encr. + Auth.)
Decryption required	Yes	No
Cipher calls (Initialization)	1	1
Cipher calls (Encryption)	$\lceil |M|/n \rceil + 1.016^{1}$	$\lceil |M|/n \rceil + 1$

[1] Applies as long as the associated data is fixed during a single session and a counter is used for the nonce.

applications [15] what may increase the popularity of OCB in the near future. Table 2 summarizes the properties of the two AE block cipher modes of operation. One benefit of GMC might be that it solely requires the encryption of the underlying block cipher whereas OCB also needs the decryption.

Our designs have been tested on a self-designed printed circuit board (PCB), which has solely been developed for high-speed authenticated encryption architectures running on an Altera Stratix IV FPGA.

6   100 Gbit/s Authenticated Encryption System Design

The AE core, described in Section 4, was developed as part of a larger FPGA-based system that will be used to encrypt data on IEEE 802.3ba Ethernet connections, allowing data rates of up to 100 Gbit/s. Designing a real system that is able to reliably process such high data-throughputs poses some formidable challenges. We have successfully developed a complete FPGA-based system working at 40 respectively 100 Gbit/s and will present both the digital part and the PCB development throughout the next sections.

6.1   FPGA Digital Design

Processing data at a rate of up to 100 Gbit/s by itself is per se rather challenging. However, transporting this amount of data to and from the processing cores is also a significant problem. In this system we have decided to aggregate plaintext data from ten separate 10 Gbit/s Ethernet links into a single 100 Gbit/s data stream. This data is then encrypted and an authentication tag is determined using the AE schemes mentioned earlier. The resulting ciphertext data stream is then transmitted over a single 100 Gbit/s Ethernet link. As illustrated in Fig. 5, the receiving path of the system works similarly in the opposite direction. The system ensures that the AE remains transparent for all 10 Gbit/s clients connected to the system.

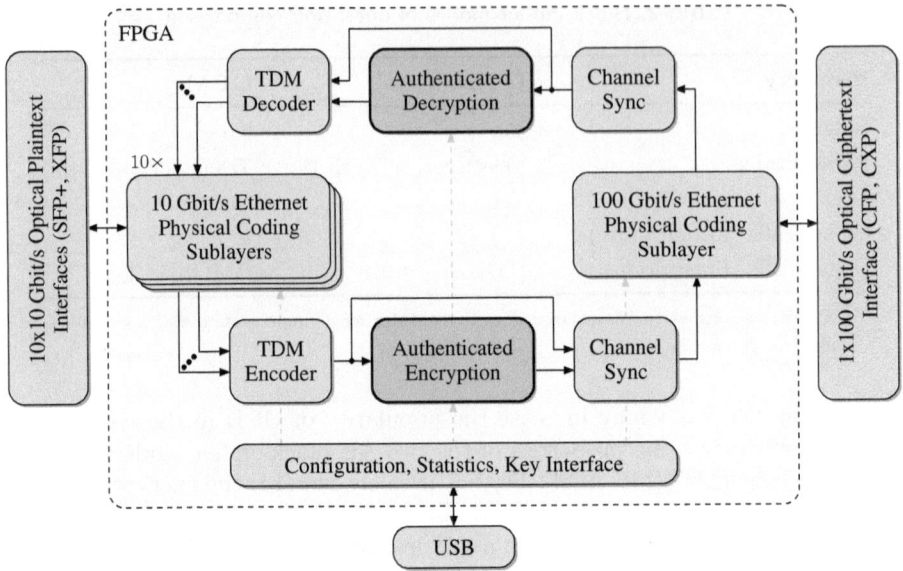

Fig. 5. Block diagram of the whole encryption system on the FPGA

Transmitting Path. The plaintext, received from the 10 Gbit/s Ethernet clients, arrives as a serial data stream. In the 10 Gbit/s Ethernet *Physical Coding Sublayer* (PCS) block, this data is parallelized and prepared for further operations. Note that in our implementation we do not require a Media Access Control (MAC) unit. Instead, we directly aggregate and encrypt the received data stream. The *TDM Encoder* collects the data from each PCS and transfers it to the *Authenticated Encryption* core, where the data is encrypted and a corresponding authentication tag is generated. The *Channel Sync* unit encapsulates the encrypted data and its authentication tag into a TDM Ethernet frame. To ensure fast resynchronization after a connection or packet loss, a synchronization frame is inserted every millisecond into the 100 Gbit/s stream. The synchronization frame is also used to transmit parameters such as the current initialization vector for GCM and the nonce for OCB. The generated frames from the *Channel Sync* block are prepared for transmission and serialized in the 100 Gbit/s PCS. The PCS uses ten 10 Gbit/s serial data streams to transmit the data.

One problem during transmission is the synchronization loss due to various effects such as electrical and optical multiplexing in the optical CFP modules and small differences in the length of electrical traces on the PCB. As a result, the serial streams may arrive at the receiver out of order. Therefore, a mechanism is required to reorder and realign the serial streams. Unique alignment markers for each stream are inserted every 100 μs to enable synchronization at the receiver.

Receiving Path. On the receiving path, the PCS deserializes the incoming 100 Gbit/s Ethernet transmission into ten 10 Gbit/s data streams. These data streams are then reordered and possible delays are compensated by utilizing the alignment markers inserted during the transmission. The system is capable of compensating up to 200 ns of delay in this way. In the next step, the now parallelized 100 Gbit/s data stream is decoded by the receiving *Channel Sync*. If a synchronization frame is detected, parameters for the AE cores are extracted from this frame and applied for the following data frames. When a TDM Ethernet frame is detected, the payload is extracted and sent to the *Authenticated Decryption* core. In the *TDM Decoder*, the decrypted data stream is distributed to the corresponding 10 Gbit/s PCS units. In addition, the calculated authentication tag from the *Authenticated Decryption* block is compared to the received one. If an authentication failure is detected, an alert flag is set. The system can be configured to react with further measures, such as purging its input data for that channel.

An important problem of the system is clock synchronization. It occurs if the clock on the receiving side of the system differs (slightly) from the transmitting side. According to the Ethernet standard, the maximum allowed clock mismatch is 100 ppm. To be able to compensate these mismatches, the system can enlarge or shrink the gap between two Ethernet frames.

System Configuration. The system on the FPGA can be configured and monitored via a *USB* connection in the development board. Individual 10 Gbit/s links can be disabled, the encryption can be turned on or off, and secret key sizes can be determined through this interface. In addition, the encryption keys are also submitted via this interface. Moreover, the same connection can be used to monitor the operation. Statistical data such as number of transmitted and received frames, status of the *Authenticated En/Decryption* blocks, or presence and link activity of SFP+, XFP, CFP, and CXP modules can be observed using the configuration interface.

Performance. Although our development board and the AE cores have been designed to support a 100 Gbit/s communication, real-world experiments have so far only been undertaken using a 40 Gbit/s ciphertext interface due to financial reasons[4]. Nevertheless, measurements of the overall system proved it to be operational at data rates up to 40 Gbit/s with all the features described above. The whole digital system showed a constant latency of 3.5 μs for all 10 Gbit/s Ethernet links when configured with OCB-Serpent. Correct transmission of frames up to a length of 16,000 Bytes was observed. The total power dissipation of the overall development board is 45 W, thereof 14 W are consumed solely by the FPGA. If configured with OCB-Serpent, the FPGA's utilization corresponds to 32 %. Note that when operating in the 40 Gbit/s configuration, only two fully unrolled encryption cores for the transmitting path and two fully unrolled decryption cores for the receiving path are required.

---

[4] The 100 Gbit/s CFP module is about eight to ten times more expensive than the 40 Gbit/s module.

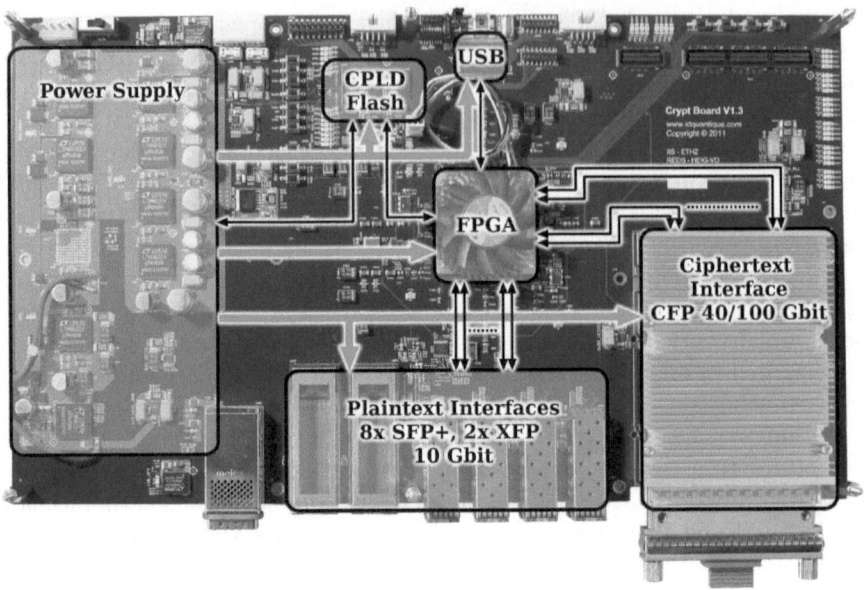

Fig. 6. Overview of the 100 Gbit/s AE development board

6.2   PCB Design

Designing such a complex system in one shot is, in our opinion, not a sound engineering strategy. While reconfiguration of the FPGA does not pose a problem at all, the design of the underlying PCB is "rather statical". Therefore, we have adopted a two-stage design process where we have developed two PCBs. The first system shown in Fig. 6, with its main features listed in Table 3, was used as a development board with all the main components and allowed us to test the basic functionality. The second PCB is the final prototype, and in addition to fixing problems detected in the first design, also adds a number of changes to meet the industrial requirements.

In this section, we will describe the main problems (power distribution and signal quality) we have encountered while designing the development system and will also explain the optimizations we have performed for a follow-up board.

Power Distribution. The first challenge in the system design is establishing the connections to the FPGA which uses a 1932-pin BGA package. While the signal connections offer a formidable challenge in terms of routing, the real problem is in power routing. The system uses in total 14 different power supplies and the main digital power supply of 0.95 V was estimated to consume as much as 48 A. The only practical solution to supply the FPGA with stable power is using several dedicated low-impedance power planes. As a result of these considerations, a 24-layer stack was designed for the development PCB. Even though we

Table 3. Main components of the development board for the 100 Gbit/s AE system

Component	Type	Description
Networking and encryption engine	Altera Stratix IV GT	FPGA model EP4S100G5F45 with high-speed transceivers, 1932-BGA
Board controller	Altera MAX II	CPLD EPM2210F256
Plaintext interfaces	8 SFP+	10 Gbit/s Ethernet interfaces. Short range (4 SFP+ prepared for Fibre Channel)
	2 XFP	10 Gbit/s Ethernet interfaces. Medium rage
Ciphertext interfaces	1 CFP	40 Gbit/s or 100 Gbit/s wavelength multiplexing four or ten 10 Gbit/s electrical streams per direction
	1 CXP	100 Gbit/s active cable using ten 10 Gbit/s fibres per direction. Short range
USB interface	Cypress EZ-USB FX2LP	For configuration, key transfer, and statistics
Power system	4 LMT4601	4-phase 0.95 V FPGA core supply. max. 48 A
	6 further switched regulators	Digital supplies and analog pre-supplies
	13 linear regulators	Analog and timing block supplies
Clocking system	7 oscillators	System clocks, Transceiver clocks, 24 ... 644 MHz
PCB	NELCO NP400-13EP	24 layers, 387 mm × 220 mm × 3 mm, 1175 components

had anticipated problems with the 0.95 V supply, measurements showed that the voltage-drop across the power plane was still too high. Therefore, we have added two additional power planes for the final design.

Power considerations have also dictated the organization of the layer stack. In the development board, high currents were concentrated on thicker low-impedance layers in the uppermost layers, close to the energy hungry components and their blocking capacitors. However, this asymmetric PCB stack could not be used for the second PCB, since a different manufacturer had to be used. As a result, half of the power layers had to be moved to the bottom for the second PCB.

Signal Quality. As expected, routing several 10 Gbit/s high-speed differential lines across a large PCB turned out to be a challenging task. In total, the development board used sixty impedance-matched differential lines and ten differential clock signals in the frequency range of 156 to 644 MHz. In the development board, these signals were routed on dedicated high-speed layers towards the bottom of the layer stack, where signal quality was not further compromised by longer via stubs. Although utmost care was taken in the design of these differential lines, actual measurements on the board revealed that the attenuation on high-speed signals was critical and problems were detected in impedance matching of the vias. We were able to reach the bit error rates specified in the IEEE 802.3ba standard by programming the transceivers on the FPGA side.

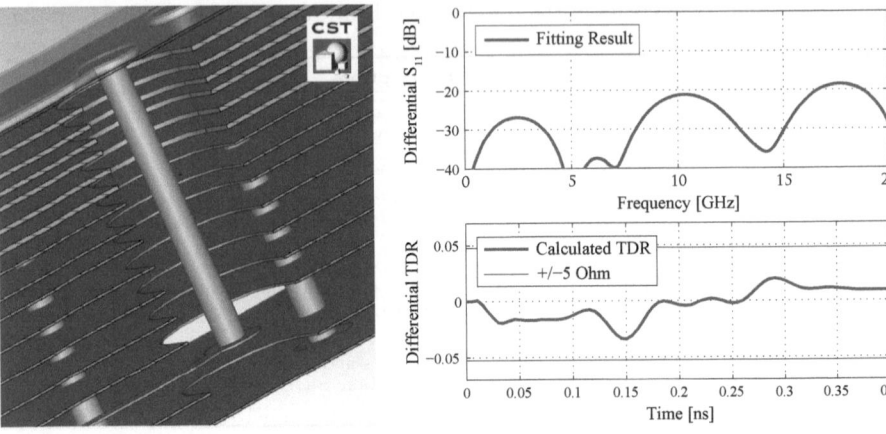

Fig. 7. 3D simulation of a differential via from an impedance-matched wire on top to a wire of layer 14 of the pad stack

Fig. 8. Reflexion coefficient S_{11} of the simulated differential via in Fig. 7 and simulated, differential time-domain reflectometry (TDR) result

As mentioned earlier, for the second PCB a new manufacturer had to be used, which necessitated a change in the layer stack. In the new layer stack, the high-speed signals are placed in the center of the stack, sandwiched between ground planes. We created a 3D model of the new layer stack using CST Microwave Studio as seen in Fig. 7. This allowed us to make detailed simulations on the behavior of differential vias and determined the best possible geometry to be used. Fig. 8 gives the result of the S_{11} parameter and simulated TDR behavior of the differential via shown in Fig. 7.

Advanced PCB. By applying a two-stage design approach, we achieved the following goals. First, the design constraints for the initial board are relaxed, allowing the board to be manufactured early in the process. Second, the development board is then actively used throughout the development of the AE core and the surrounding system, allowing real measurements on a representative system. These in turn were used to identify problems in the development board and has guided the design of the final PCB. In addition to the weaknesses detected in the first PCB design, it was decided to make the following changes to meet industrial constraints:

– Added two additional layers to improve power distribution.
– Moved the secret key port from USB to PCIe to improve throughput.
– Removed the CXP active cable interface which was deemed to be unnecessary for the application.
– Replaced the two XFP modules by two SFP+ modules.
– Added electronic dispersion compensation (EDC) to six of the SFP+ modules.

– Adapted the dimensions of the PCB to better comply with the requirements of the industrial partner.

However, our two-stage approach also had some drawbacks. The design parameters for such complex PCB systems are not standardized, and most of these parameters need to be negotiated with the PCB manufacturer directly. If for some reason, the PCB manufacturer has to be changed, it is likely that the chosen parameters can not be reused, necessitating time-consuming re-design work. In our case, we were forced to change manufacturers as the initial manufacturer filed for bankruptcy. It proved to be quite difficult to find an alternative PCB manufacturer, 12 out of the 14 companies we have contacted refused the design due to its high complexity.

The advanced PCB now has 26 layers and is 3.3 mm thick. This thickness posed additional challenges for drill holes. The maximum practical aspect ratio (thickness/diameter) for drill holes is around 16:1. Therefore, the thickness of the board directly determines the minimum diameter of the vias that can be used. Since smaller vias are required for impedance matching of 10 Gbit/s differential lines, a delicate balancing act is necessary to reconcile the demands of signal quality on one hand and safe manufacturability on the other hand.

As a consequence of both the increased number of layers, and the change of the manufacturer, we were no longer able to use vias that were 0.22 mm in diameter, but had to adjust the minimal via size to 0.25 mm. This change alone required significant re-design effort.

7   Conclusion

In this work, we described a hardware architecture for high-speed authenticated encryption (AE) using block ciphers on FPGAs, based on *alternative* cryptographic primitives. Our design operates in the offset codebook (OCB3) mode of operation and contains four parallel Serpent block cipher cores for the encryption part in order to achieve the desired data rates according to IEEE 802.3ba. The OCB3-Serpent architecture reaches a throughput of 141 Gbit/s and thus, outperforms all GCM-AES implementations on FPGAs available to date. Although OCB3 has not (yet) been approved by the NIST, its structure (as well as that of Serpent) is definitely suitable for high-throughput implementations as shown during this work. Moreover, we present a custom-designed printed circuit board for the Stratix IV FPGA, which allows us to use the presented AE schemes in real-world applications processing data with up to 100 Gbit/s.

Acknowledgement. This work is part of the QCRYPT project, which is evaluated by the Swiss National Science Foundation and financed by the Swiss Confederation with funding via Nano-Tera.ch.

The authors would like to thank the entire team from the Microelectronics Design Center at the ETH Zurich for their help during the development of the printed circuit board as well as Christian Pendl from Graz University of Technology for his contributions to the digital part of the system.

References

1. IEEE Standard for Information technology-Telecommunications and information exchange between systems - Local and metropolitan area networks - Specific requirements Part 3: Carrier Sense Multiple Access with Collision Detection (CSMA/CD) Access Method and Physical Layer Specifications Amendment 4: Media Access Control Parameters, Physical Layers, and Management Parameters for 40 Gb/s and 100 Gb/s Operation. IEEE Std 802.3ba-2010 (Amendment to IEEE Standard 802.3-2008), pp. 1–457 (22 2010)
2. Anderson, R., Biham, E., Knudsen, L.: Serpent: A Proposal for the Advanced Encryption Standard. In: Proceedings of the First AES Candidate Conference. National Institute of Standard and Technology, Ventura (1998)
3. Biham, E., Anderson, R., Knudsen, L.R.: Serpent: A New Block Cipher Proposal. In: Vaudenay, S. (ed.) FSE 1998. LNCS, vol. 1372, pp. 222–238. Springer, Heidelberg (1998)
4. Crenne, J., Cotret, P., Gogniat, G., Tessier, R., Diguet, J.P.: Efficient Key-Dependent Message Authentication in Reconfigurable Hardware. In: 2011 International Conference on Field-Programmable Technology (FPT), pp. 1–6 (December 2011)
5. Dworkin, M.: Recommendations for Block Cipher Modes of Operation: The CCM Mode for Authentication and Confidentiality. Tech. rep., NIST (2004)
6. Dworkin, M.: Recommendations for Block Cipher Modes of Operation: Galois/Counter Mode (GCM) and GMAC. Tech. rep., NIST (2007)
7. Henzen, L., Fichtner, W.: FPGA Parallel-Pipelined AES-GCM Core for 100G Ethernet Applications. In: 2010 Proceedings of the ESSCIRC, pp. 202–205 (September 2010)
8. IEEE Std 802.11-2007: IEEE Standard for Information Technology - Telecommunications and Information Exchange Between Systems - Local and Metropolitan Area Networks - Specific Requirements - Part 11: Wireless LAN Medium Access Control (MAC) and Physical Layer (PHY) Specifications (June 2007)
9. Jutla, C.S.: Encryption Modes with Almost Free Message Integrity. In: Pfitzmann, B. (ed.) EUROCRYPT 2001. LNCS, vol. 2045, pp. 529–544. Springer, Heidelberg (2001)
10. Krovetz, T., Rogaway, P.: The Software Performance of Authenticated-Encryption Modes. In: Joux, A. (ed.) FSE 2011. LNCS, vol. 6733, pp. 306–327. Springer, Heidelberg (2011)
11. McGrew, D.A., Viega, J.: The Galois/Counter Mode of Operation (GCM). NIST Modes Operation Symmetric Key Block Ciphers (2005)
12. Muehlberghuber, M., Keller, C., Felber, N., Pendl, C.: 100 Gbit/s Authenticated Encryption Based on Quantum Key Distribution. In: 2012 IEEE/IFIP 20th International Conference on VLSI and System-on-Chip (VLSI-SoC), pp. 123–128 (October 2012)
13. Nechvatal, J., Barker, E., Bassham, L., Burr, W., Dworkin, M., Foti, J., Roback, E.: Report on the Development of the Advanced Encryption Standard (AES). Tech. rep., National Institute of Standards and Technology, NIST (2000)
14. NIST: Advanced Encryption Standard (AES) (FIPS PUB 197). National Institute of Standards and Technology (November 2001)
15. Rogaway, P.: OCB Free Licenses (2013), http://www.cs.ucdavis.edu/~rogaway/ocb/license.htm (accessed March 05, 2013)

16. Rogaway, P., Bellare, M., Black, J., Krovetz, T.: OCB: A Block-Cipher Mode of Operation for Efficient Authenticated Encryption. In: ACM Conference on Computer and Communications Security, pp. 196–205 (2001)
17. Viega, J., Mcgrew, D.: The Use of Galois/Counter Mode (GCM) in IPsec Encapsulating Security Payload (ESP). RFC 4106 (Proposed Standard) (June 2005)
18. Wang, J., Shou, G., Hu, Y., Guo, Z.: High-speed architectures for GHASH based on efficient bit-parallel multipliers. In: 2010 IEEE International Conference on Wireless Communications, Networking and Information Security (WCNIS), pp. 582–586 (2010)
19. Whiting, D., Housley, R., Ferguson, N.: Counter with CBC-MAC (CCM). RFC 3610 (Informational) (September 2003)
20. Zhou, G., Michalik, H., Hinsenkamp, L.: Improving Throughput of AES-GCM with Pipelined Karatsuba Multipliers on FPGAs. In: Becker, J., Woods, R., Athanas, P., Morgan, F. (eds.) ARC 2009. LNCS, vol. 5453, pp. 193–203. Springer, Heidelberg (2009)

A   OCB Algorithms

Algorithm 2 lists the calculation of the table values $L[..]$ required during the OCB mode of operation. The *double*-procedure is defined according to:

$$double(X) = (X \ll 1) \oplus (msb(X) \cdot 0x87),$$

where $msb(X)$ represents the most significant bit of X using binary representation.

Algorithm 2. Table value calculations

Input: Cipher key K, Number of message blocks m
Output: $Setup(K, m)$
 1: $L_* \leftarrow E_K(0^{128})$
 2: $L_\$ \leftarrow double(L_*)$
 3: $L[0] \leftarrow double(L_\$)$
 4: **for** $i = 1$ to $\lfloor log_2(m) \rfloor$ **do**
 5:     $L[i] \leftarrow double(L[i-1])$
 6: **end for**
 7: **return** $L_*, L_\$, L[0] \ldots L[\lfloor log_2(m) \rfloor]$

The initial offset Δ is determined according to Algorithm 3. $X \ll i$ denotes a left shift operation of X by i bits.

Algorithm 3. Initial offset (Δ) calculation

Input: Nonce N, Message block length n, Cipher key K
Output: $Init(N, n, K)$
  1: $Bottom \leftarrow N \wedge 1^6$ ▷ $Bottom$ = Least six LSBs of N
  2: $Nonce \leftarrow 0^{127-|N|}||1||N$
  3: $Top \leftarrow (1^{122}||0^6) \wedge Nonce$ ▷ Zeroing out least six LSBs of $Nonce$
  4: $Ktop \leftarrow E_K(Top)$
  5: $Stretch \leftarrow Ktop||(Ktop \oplus (Ktop \ll 8))$
  6: $\Delta \leftarrow (Stretch \ll Bottom) \wedge 1^n$ ▷ Use first n bits of $Stretch \ll Bottom$
  7: **return** Δ

Algorithm 4 describes the calculation of $Hash_K(A)$. Since the *Setup* procedure already gets called during the actual encryption process of OCB (cf. Algorithm 1, line 4), line 3 in Algorithm 4 can be omitted as long as the table values $L[..]$ are globally available.

Algorithm 4. Authentication hash ($Hash_K(A)$) calculation

Input: Associated data A, Associated data block length q, Cipher key K
Output: $Hash_K(A)$
  1: $\{A_1, \ldots, A_p, A_*\} \leftarrow A$, with $|A_i| = q$ and $|A_*| < q$
  2: $Sum \leftarrow 0^{128}$; $\Delta \leftarrow 0^{128}$
  3: $L_*, L[0] \ldots L[\lfloor log_2(p) \rfloor] \leftarrow Setup(K, p)$
  4: **for** $i = 1$ to p **do**
  5:      $\Delta \leftarrow \Delta \oplus L[ntz(i)]$ ▷ $Inc_i(\Delta)$
  6:      $Sum \leftarrow Sum \oplus E_K(A_i \oplus \Delta)$
  7: **end for**
  8: **if** $A_* \neq \varnothing$ **then**
  9:      $\Delta \leftarrow \Delta \oplus L_*$ ▷ $Inc_*(\Delta)$
10:      $Sum \leftarrow Sum \oplus E_K(A_*10^* \oplus \Delta)$, with
                 $A_*10^* = A_*||1||0 \ldots 0$, such that $|A_*10^*| = q$
11: **end if**
12: **return** Sum

A Smart Memory Accelerated Computed Tomography Parallel Backprojection

Qiuling Zhu, Larry Pileggi, and Franz Franchetti

Department of Electrical and Computer Engineering
Carnegie Mellon University, Pittsburgh, PA, USA
qiulingz@andrew.cmu.edu

Abstract. As nanoscale lithography challenges mandate greater pattern regularity and commonality for logic and memory circuits, new opportunities are created to affordably synthesize more powerful smart memory blocks for specific applications. Leveraging the ability to embed logic inside the memory block boundary, we demonstrate the synthesis of smart memory architectures that exploits the inherent memory address patterns of the backprojection algorithm to enable efficient parallel image reconstruction at minimum hardware overhead. An end-to-end design framework in sub-20nm CMOS technologies was constructed for the physical synthesis of smart memories and evaluation of the huge design space. Our experimental results show that customizing memory for the computerized tomography (CT) parallel backprojection can achieve more than 30% area and power savings while offering significant performance improvements with marginal sacrifice of image accuracy.

Keywords: Smart Memory, Logic and Memory Synthesis, Computed Tomography, Parallel Backprojection.

1   Introduction

Computationally intensive algorithms in medical image processing (e.g., computerized tomography (CT)) require rapid processing of large amounts of data and often rely on hardware acceleration [1–3]. Inherent parallelism in the algorithms is exploited to achieve the required performance by increasing the number of parallel functional units at a cost of power and area. The overall performance is often defined by the limited bandwidth of the on-chip memory as well as the high cost of memory access.

One approach to address these challenges is to optimize the on-chip memory organization by constructing a customized smart memory module that is optimized for a particular function for higher performance and/or energy efficiency [4, 5]. However, such customization is generally unaffordable for an application-specific IC embedded memory for which cost dictates that it is "compiled" from a set of SRAM hard IP components (e.g., physical implementations of bitcells and peripheral circuits). Such memory compilation limits the possibility of application-specific customization and hinders the system design space exploration.

A. Burg et al. (Eds.): VLSI-SoC 2012, IFIP AICT 418, pp. 21–44, 2013.
© IFIP International Federation for Information Processing 2013

Recent studies of sub-20nm CMOS design indicate that memory and logic circuits can be implemented together using a small set of well-characterized pattern constructs [6, 7]. Our early silicon experiments in a sub-20nm commercial SOI CMOS process demonstrate that this construct-based design enables logic and bitcells to be placed in a much closer proximity to each other without yield or hotspots pattern concerns. While such patterning appears to be more restrictive to accommodate the physical realities of sub-20nm CMOS, the ability to make the patterns the only required hard IP allows us to efficiently and affordably customize the SRAM blocks. More importantly, it enables the synthesis (not just compilation) of customized memory blocks with user control of flexible SRAM architectures and facilitate *smart memory compilation.*

To efficiently leverage this new technology, however, algorithms and hardware architectures need to be revised. In this paper we revisit the well-known Shepp and Logan's backprojection algorithm that is widely used in the CT image reconstruction [3]. It is observed that in the parallel implementation of the algorithm, the memory address differences are fairly small for adjacent projection angles and adjacent pixels. We exploit this property via a customized memory structure that could feed in-parallel running image processing engines (IPEs) with a large amount of required projection data in one clock cycle. The implementation is realized by embedding "intelligent" functionality into the traditional interleaved memory organization and allow multiple memory sub-banks to share the memory periphery. Novel periphery-sharing smart memory strategies are explored, and an efficient parallel-pipeline backprojection architecture is proposed. We further construct a smart memory design framework that provides the end user with finer control of the customized SRAM architecture parameters, thus enabling automatic generation of the specified implementation. Physical implementations were carried out in a commercial sub-20nm SOI CMOS process. Our results indicate that there is more than 40% area savings and 30% power savings while providing significant performance improvements. The marginal impact on accuracy is minimized with appropriate constraints on the algorithm.

Related Work. In other related work various fast approaches have been proposed to improve the backprojection implementation [2, 8, 9, 3]. As pointed out in [3], these approaches may be classified into three categories; namely, algorithmic improvement, dedicated hardware, and parallel processing. However, this paper shows that it is possible to combine these three aspects to deliver a more efficient backprojection architecture by taking advantage of the availability of smart memory synthesis. Our approach optimizes the parallel backprojection architecture, especially the on-chip memory architecture, by exploiting the inherent memory address pattern that has not been previously explored.

2   Background

Filtered backprojection is the most commonly used approach for image reconstruction from parallel-beam projection data. Before analyzing the inherent

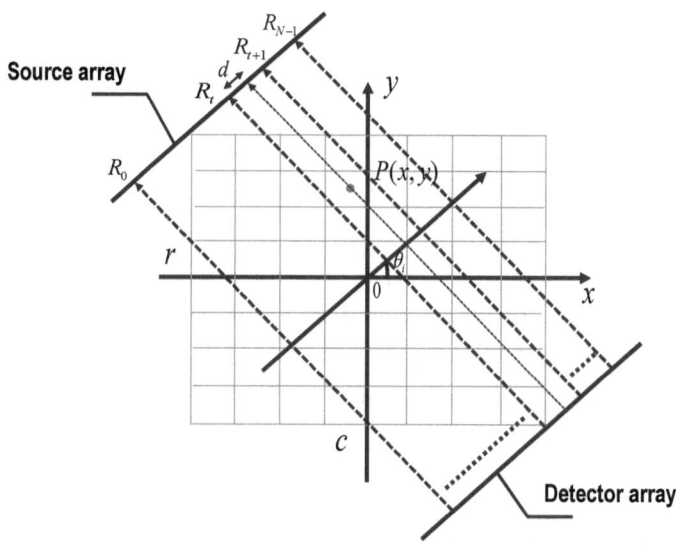

Fig. 1. Illustration of Parallel-Beam Projection: The object to be scanned is placed between the evenly spaced array of an unidirectional X-ray source and the detector. Radiation beams from the X-ray source pass through the object and are measured at the detector, forms the projections of the image.

memory access pattern and building the corresponding customized memory architecture, in this section we will first introduce the parallel-beam CT scanning system and the commonly used backprojection algorithm.

2.1 CT Scanning Method

Tomography is a non-invasive imaging technique allowing for the visualization of the internal structures of an object. Tomography has found widespread applications in many scientific fields, including physics, chemistry, astronomy, geophysics, and medicine. A parallel-beam CT scanning system uses an array of equally spaced unidirectional sources of focused X-ray beams. The object to be scanned is placed between the sources array and the detector. Radiation beams from the source pass through the object and are measured at the detector (see Fig. 1). A complete set of projections is obtained by rotating the arrays and taking measurements for different angles over 180°, forming the Radon transform of the image (i.e., projection data), and it contains information needed for the reconstruction of an image. A set of values given by all detectors in the array comprises a one-dimensional projection data. The inverse of the projection data allows to reconstruct the tomographic images (i.e., backprojection) [10, 1]. The Radon transform and its inverse provide the mathematical basis for reconstructing tomographic images from the measured projection data.

2.2 Shepp and Logan Backprojection Algorithm

The most widely known reconstruction-from-projections test image is the Shepp-Logan phantom. Introduced in 1974 it is still in common use today as a reference image for reconstruction algorithms. The Shepp and Logan backprojection algorithm is the most well-known backprojection algortithm [3, 11]. In the conventional Shepp and Logan backprojection algorithm, for each pixel, P, located at (x, y), and each projection angle θ_i, the first step in backprojection is to locate the pixel in an appropriate beam (ray). If the center of P is not on a ray, the distance (d) to its adjacent rays is calculated and the contribution from the adjacent rays to the pixel (Q_p) is computed according to the linear interpolation equation (1), assuming that pixel is enclosed by the t_{th} and $(t + 1)_{th}$th rays,

$$Q_p(x, y, \theta_i) = R_t + (d/L) \cdot (R_{t+1} - R_t), \tag{1}$$

where R_t is the value of t_{th} ray, d is the interpolation distance, and L is the ray interval. Q_p represents the contribution of the projection of angle θ_i to the current pixel value.

In the above equation, the address t to the projection data memory and the interpolation distance d are computed as follows (assuming the target image has the dimension size of $r \times c$):

$$t_{x,y,\theta_i} = \left(x - \frac{r}{2}\right) \cdot \cos\theta_i - \left(y - \frac{c}{2}\right) \cdot \sin\theta_i + t_{\text{offset}}. \tag{2}$$

and the interpolation distance d is calculated as follows:

$$d = t(\theta) - \lfloor t(\theta) \rfloor. \tag{3}$$

Existing Algorithm Optimization. The above procedures, locating and interpolation, are to be repeated for every pixel and for every projection angle. However, there exists computational redundancy that can be explored to save the operations in the iterations. To do this, $2D$ Shepp and Logan algorithm exploits the property of constant difference of address t for those pixels on the same row or column. Considering two adjacent pixels located at (x, y) and $(x + 1, y)$, and backprojection angle θ, we can calculate the addresses to the projection memory for the two pixels based on (2). And the difference of their addresses, $t_{x+1,y,\theta_i} - t_{x,y,\theta_i}$ is equal to $cos(\theta)$, which is a constant for a given θ. Let δt_x denotes the constant difference along the x direction. Then, in the $2 - D$ Shepp and Logan algorithm, instead of evaluating equation (2) for every pixel, it simply adds a constant of $cos(\theta)$ to the previous adjacent address index (t_{x,y,θ_i}) to generate the new address index (t_{x+1,y,θ_i}). The same rule can be applied to the y direction to calculate the address index $(t_{x,y+1,\theta_i})$ by adding the constant difference δt_y of $sin(\theta)$ to (t_{x,y,θ_i}).

3   Memory Address Pattern Analysis

This paper moves one step forward by taking advantage of these constant address differences of δt_x and δt_y that exist in the conventional Shepp and Logan

Backprojection algorithm, to simplify not only the address calculation but also the underlying memory hardware. Furthermore, we will demonstrate that the address differences when the projection angle θ changes are also within a very small and predictable range that could be also exploited to further optimize the hardware memory design.

3.1   Address Difference for Adjacent Projections

For each pixel (x, y) and each projection angle (θ_i), the beam index t_{x,y,θ_i} (i.e., address to the projection memory) is already shown as in (2). To illustrate the inherent address patterns that were hidden in the algorithm, we show the address to the next projection of angle θ_{i+1} in (4):

$$t_{x,y,\theta_{i+1}} = \left(x - \frac{r}{2}\right) \cdot \cos\left(\theta_{i+1}\right) - \left(y - \frac{c}{2}\right) \cdot \sin\left(\theta_{i+1}\right) + t_{\text{offset}}. \tag{4}$$

The address difference (δt_1) between (2) and (4) could be as

$$\delta t_1 = \left(x - \frac{r}{2}\right) \cdot \delta cos_{\theta_i} + \left(\frac{c}{2} - y\right) \cdot \delta sin_{\theta_i}, \tag{5}$$

with $\delta cos_{\theta_i} = \cos(\theta_{i+1}) - \cos(\theta_i)$ and $\delta sin_{\theta_i} = \sin(\theta_{i+1}) - \sin(\theta_i)$. δcos_{θ_i} can be rewritten as:

$$\delta cos_{\theta_i} = \cos\theta_{i+1} - \cos\theta_i = -2\sin\frac{\theta_{i+1} + \theta_i}{2}\sin\frac{\theta_{i+1} - \theta_i}{2} \tag{6}$$

For $\theta_i = \frac{2\pi i}{N}$, $\frac{\theta_{i+1} - \theta_i}{2}$ is the constant π/N, so δcos_{θ_i} is simplified as

$$\delta cos_{\theta_i} = -2\sin\left(\frac{\pi}{N}\right)\sin\left(\frac{\pi(2i + 1)}{N}\right) \tag{7}$$

Similarly, we have:

$$\delta sin_{\theta_i} = 2\sin\left(\frac{\pi}{N}\right)\cos\left(\frac{\pi(2i + 1)}{N}\right) \tag{8}$$

So, (5) can be written as:

$$\delta t_1 = \left(x - \frac{r}{2}\right) \cdot \left(-2\sin(\frac{\pi}{N})\sin\frac{\pi(2i + 1)}{N}\right) + \left(\frac{c}{2} - y\right) \cdot \left(2\sin(\frac{\pi}{N})\cos\frac{\pi(2i + 1)}{N}\right) \tag{9}$$

Therefore, using trigonometric identities, we can compute the bound on (5) as follows:

$$|\delta t_1| \leq |2 \cdot \sin\left(\frac{\pi}{N}\right) \cdot \frac{r}{2} \cdot \left(\cos\left(\frac{\pi(2i + 1)}{N}\right) - \sin\left(\frac{\pi(2i + 1)}{N}\right)\right)|. \tag{10}$$

(10) has a maximum bound of $\sqrt{2}\pi \cdot \frac{r}{N}$ for relatively large N.

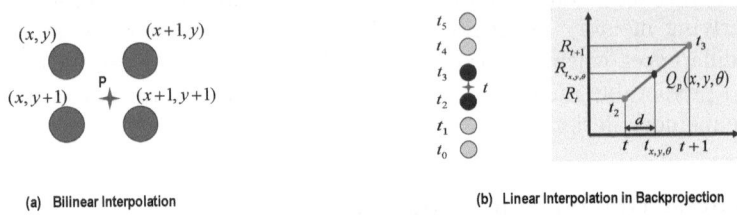

(a) Bilinear Interpolation

(b) Linear Interpolation in Backprojection

Fig. 2. Interpolation in CT Backprojection

Here we assuming $r = c$ is the dimension size of a square image and N is the number of projection angles. It is shown that δt_{θ_1} is restricted in a very limited range when the ratio of r and N is relatively small. For example, δt_{θ_1} must be less than 1 when $\frac{r}{N} \leq \frac{1}{8}$.

This observation can easily extend to the scenario of computing the contribution of consecutive k projection angles to the same pixel (x, y). In this situation, the address differences will be accumulated and the resulting accumulating address difference between the next k projection memory of angle θ_k and the first memory of angle θ_1 for the same pixel $P(x, y)$ will increase proportionally to k:

$$|\delta t_k| = |t_{x,y,\theta_{i+k}} - t_{x,y,\theta_i}| \leq \sqrt{2}\pi \cdot \frac{r}{N} \cdot k \cdot \approx 4.44 \cdot \frac{r}{N} \cdot k. \qquad (11)$$

For certain value of k, δt_k will still be within a very small value.

3.2 Address Difference for Adjacent Pixels

In the above section, we have derived the beam index differences for a fixed pixel when projection angles increment. Next, we will show that the address differences when both pixel coordinate and projection angle increment are also bounded by a limited range.

For demonstration purpose, we define the problem as to reconstruct four neighborhood pixels in parallel, that is, (x, y), $(x + 1, y)$, $(x, y + 1)$, $(x + 1, y + 1)$. We will encounter this problem for parallel image reconstruction. For example, Fig. 2 shows the example to compute four neighborhood pixels, (x, y), $(x + 1, y)$, $(x, y + 1)$, $(x + 1, y + 1)$, in parallel. The similar problem could also happen in a higher-level interpolation, that is, the calculation of the non-existing pixel P requires to compute its four neighborhood pixels first and apply a bilinear interpolation afterwards.

We denote the address of the first pixel (x, y) in the first projection memory θ_i as the reference address (t_{x,y,θ_i}). Then, for other three pixels, $(x + 1, y)$, $(x, y+1)$, $(x+1, y+1)$ in the same projection memory of θ_i, their address differences from t_{x,y,θ_i}, can be estimated as shown in (12), (13), (14):

$$|t_{x+1,y,\theta_i} - t_{x,y,\theta_i}| = |\cos(\theta_i)| \leq 1 \qquad (12)$$

$$|t_{x,y+1,\theta_i} - t_{x,y,\theta_i}| = |\sin(\theta_i)| \leq 1 \tag{13}$$

$$|t_{x+1,y+1,\theta_i} - t_{x,y,\theta_i}| = |\cos(\theta_i) + \sin(\theta_i)| \leq \sqrt{2} \tag{14}$$

It can be observed that all the shown three address differences are all in a very small range. For the same four pixels, let's now analyze their addresses to the next adjacent projection memory of angle θ_{i+1}. For the first pixel located at (x, y), its address difference from t_{x,y,θ_i} has been calculated in (10) and here we repeated it in (15):

$$|t_{x,y,\theta_{i+1}} - t_{x,y,\theta_i}| = |\delta t_{\theta_1}| \leq \sqrt{2}\pi \cdot \frac{r}{N} \tag{15}$$

Similarly, for the other three pixels, we show their address differences from t_{x,y,θ_i} in (16), (17), (18) respectively:

$$|t_{x+1,y,\theta_{i+1}} - t_{x,y,\theta_i}| = |\cos(\theta_i) + \delta t_{\theta_1}| \leq 1 + \sqrt{2}\pi \cdot \frac{r}{N} \tag{16}$$

$$|t_{x,y+1,\theta_{i+1}} - t_{x,y,\theta_i}| = |\sin(\theta_i) + \delta t_{\theta_1}| \leq 1 + \sqrt{2}\pi \cdot \frac{r}{N} \tag{17}$$

$$|t_{x+1,y+1,\theta_{i+1}} - t_{x,y,\theta_i}| = |\cos(\theta_i) + \sin(\theta_i) + \delta t_{\theta_1}| \leq \sqrt{2} + \sqrt{2}\pi \cdot \frac{r}{N} \tag{18}$$

It can be observed that all of these memory addresses in adjacent projection angles i and $i + 1$ are all very close to reference address t_{x,y,θ_i}. (18) presents the largest possible address distance among them. This is because the pixel to compute in (18) is located at $(x + 1, y + 1)$, and it changes from the pixel $p(x, y)$ in both x dimension and y dimension while pixels $p(x+1, y)$ and $p(x, y+1)$ only changes from the pixel $p(x, y)$ in either x dimension or y dimension. Therefore, the address difference between $t_{x+1,y+1,\theta_{i+1}}$ and t_{x,y,θ_i} shown in (18) is relatively larger than the other address differences from (15) to (17).

We could extend the observation to the addresses of these four pixels in the next adjacent k projection memories, that is, for projection angles from θ_i and θ_{i+k}. We can easily prove that the involved addresses are also very close to t_{x,y,θ_i} for the required k, and the maximum possible address difference to t_{x,y,θ_i} is introduced by the last pixel $(x + 1, y + 1)$ in the last projection memory θ_{i+k},

$$|\delta t_{max}| = |t_{x+1,y+1,\theta_{i+k}} - t_{x,y,\theta_i}| = |\cos\theta_i + \sin\theta_i + k \cdot \delta t_1| \tag{19}$$

(19) has the maximum value of $\sqrt{2} + 4.44 \cdot \frac{r}{N} \cdot k$ and it is limited to small range, e.g., the value must be less than four when $\frac{r}{N} \leq \frac{1}{8}$ and $k = 4$.

The basic idea is, since the address differences for adjacent projections angles and adjacent pixels are small, these addresses will activate the same or adjacent wordlines when such memories are located horizontally in parallel with each other. Such particular memory address pattern leads to opportunities to share the memory decoder among these memories by programming extra "intelligent" logic functionalities into the memory periphery.

4   Backprojection Smart Memory Design

In this section, we describe our approach to optimize the memory organization and backprojection architecture based on the observed memory access patterns.

4.1   Interpolation Memory

As we mentioned, linear interpolation is required if the location of a pixel in a specific view in not on a ray. As shown in Fig. 2 (b), if the beam index in a projection memory, t, is not an integer and located in between $[t_2,t_3]$, then the neighborhood pixels t_2 and t_3 will be accessed and an linear interpolation will be performed to compute the required pixel value t. To improve the processing speed, the neighborhood pixels t_2 and t_3 need to be accessed from the memory in one clock cycle. For the single port memory design, this requires to divide the memory into two different memory banks. Therefore, to run two adjacent backprojections in parallel, it requires to implement two separate projection memories, and each memory is divided into two memory banks. Similarly, to run more adjacent backprojections in parallel, it requires to implement more multi-banking projection memories. However, we will show that it is possible to significantly optimize the hardware implementation of such multi-banking memory system if the discussed memory address patterns are well exploited.

4.2   Consecutive Access Memory

We have discussed that linear interpolation operation requires to access two nearest neighborhood pixels from the projection memory in one clock cycle. We would like to extend this operation to access more than two consecutive pixels from the memory in one clock cycle (i.e., multiple consecutive access memory). We will show later in section 4.4 that such multiple neighborhood pixels access will be required to our smart memory design.

We will first introduce a smart memory structure which can output arbitrary number of adjacent memory entries at arbitrary position in one clock cycle. As we have mentioned, this is traditionally accomplished by distributing data across multiple memory banks so that for any consecutive access all data elements are retrieved from different banks without conflicts. Using multiple SRAM banks incurs high overhead since every memory bank requires its own decoder logic. In our previous work [12], we have proposed a *rectangular-access smart memory* which is able to output an arbitrary rectangular block in a 2D data array. Its $1D$ simplified version, called $1D$ *Consecutive Access Memory*, can be used to output consecutive elements from a $1D$ data array.

We exploit the fact that we always read a constant number of consecutive elements per cycle for each operation. The core observation is that after address decoding, the activated wordlines of all memory banks are always adjacent to each other. Based on that, it's possible to optimize the multi-banking memory system to save the periphery overhead. We employ a customized multi-banking

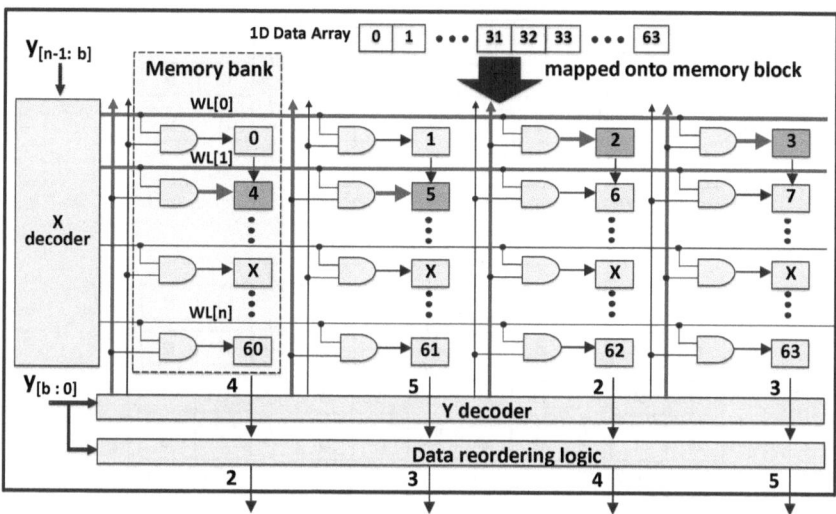

Fig. 3. Consecutive Access Memory. As the basic memory structure in the paper, our customized memory can output consecutive memory entries in one clock cycle and allows parallel memory banks to share the x-decoder.

SRAM design topology [13], which provides around 50% area and power savings compared with the traditional multi-banking memory design. We define the functionality of memory to support one-clock-cycle access of 2^b data points from a 2^n size data array. We build a parameterized memory which is divided into 2^b memory banks and they are located vertically parallel to each other. To control the memory block aspect ratio, we let each word of a memory bank holds 2^c data points. Fig. 3 shows the organization of the memory block when $n = 6$, $b = 2$, $c = 1$. The main idea is to let 2^b memory banks in each memory block share a modified X-decoder by using the same method described in [13]. The X-decoder is specifically designed to activate two adjacent wordlines simultaneously. That is, when one block wordline is asserted, the next block wordline is also asserted by the OR gate operation of every two adjacent wordline signals. Another Y-decoder is used to select one of the two activated wordlines for each memory bank with the AND operations. Each memory bank word holds 2^c data points but each time only one data point of them is required. A column MUX is designed to select one data element for each memory bank and the column MUX is controlled by the lower $(b + c)$ bits of address y ($y_{[b+c-1:0]}$).

As shown in Fig. 3, both the first wordline ($WL[0]$) and the second wordline ($WL[1]$) are initially activated by X-decoder but Y-decoder further selects the $WL[1]$ for the first two memory banks and $WL[0]$ for the last two memory banks with the additional AND operations. After the column MUX, this memory block outputs data series of '$4-5-2-3$', which are then reordered to be '$2-3-4-5$'. So with some simple logic for data reordering, the smart memory outputs the required 2^b data points in order simultaneously. The distribution of address bits

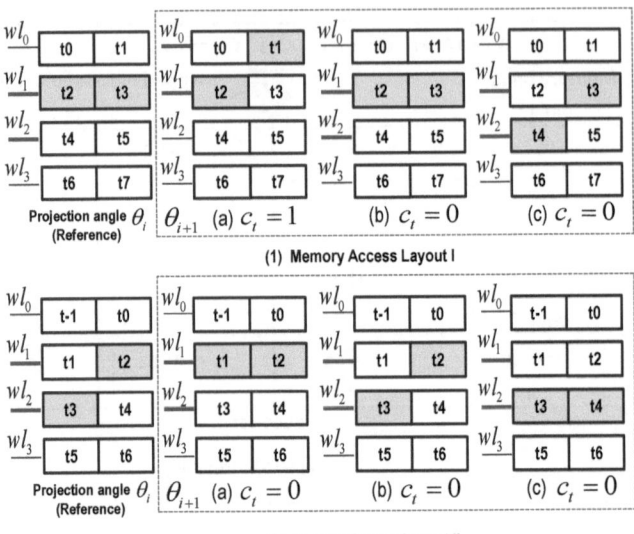

Fig. 4. Data Layout in Adjacent Two Projection Memories. If t_2 and t_3 are required in the first reference memory of the projection θ_i, then beam pixel required in the next memory of projection θ_{i+1} has three possible locations, that is, $[t_1, t_2]$, $[t_2, t_3]$ or $[t_3, t_4]$.

to each memory component is parameterized. By specify these parameters, the resulting memory architecture can be precisely determined. Therefore, we can program the smart memory at the RTL level. Compared with the conventional multi-banking memory design, the amount of memory bank periphery circuits is reduced from 2^b to 1. As is observed in Fig. 3, the resulting memory architecture has the embedded logic gates (e.g. the AND gates) which is tightly integrated with the memory cells, and each logic gate communicates with its local memory cells.

This consecutive access memory serves as the basic memory structure in our method. However, this smart memory structure could be further optimized if provided more knowledge from a particular application. In the rest of paper, we will propose more advanced memory sharing strategies to further optimize the consecutive access memory based on the observed memory access patterns in the backprojection algorithm.

4.3 Decoder-mux and Output-mux

As a simple illustration, in Fig. 4 we show the physical data layout in our consecutive access memory. If the address of projection θ_i is located in between t_2 and t_3 (denoted by $[t_2, t_3]$), then in our previous discussed consecutive access memory design, t_2 and t_3 should either be located in the same wordline or split into two separate wordlines, as shown in the first memory array in Fig. 4 (a) and Fig. 4 (b) respectively. In both situations, two wordlines, wl_1 and wl_2, are activated simultaneously. From the analysis of equation (10), we have derived that

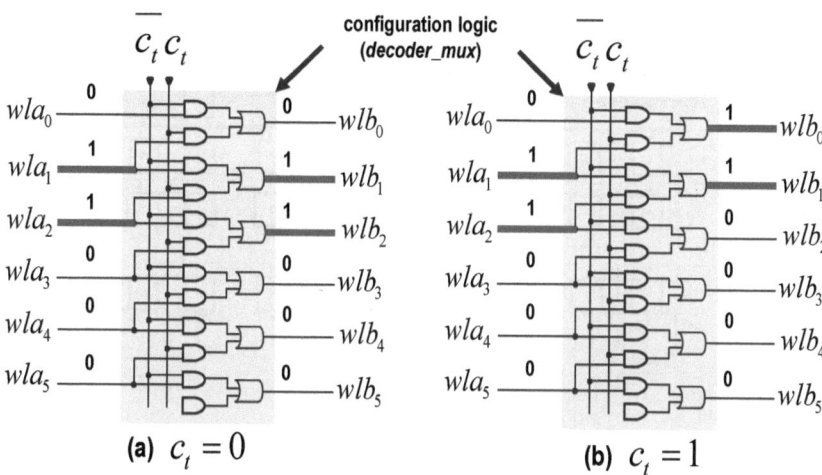

Fig. 5. Decoder-MUX. The wordlines of the first memory (wla_i) are configured to generate the wordlines for the next memory (wlb_i), so that the decoder of the latter memory could be eliminated.

the address difference of the two adjacent memories (δt_{θ_1}) is less than one when $\frac{r}{N} \leq \frac{1}{8}$. This implies that the two adjacent memory addresses after rounding must be either the same or adjacent to each other. Then for the addressed beam index of the next projection memory of angle θ_{i+1}, it will has only three possible locations, that is, $[t_1, t_2]$, $[t_2, t_3]$ or $[t_3, t_4]$, as illustrated in the next three memory layouts of Fig. 4 (1) and Fig. 4 (2). In the illustration we also highlight the corresponding active wordlines if implemented in the consecutive access memory. It's seen that if the active wordlines for the first memory are wl_1 and wl_2, then in the next memory, the active wordlines must be the same in most situations. The only exception is to access t_2 and t_3 from the first projection memory but to access t_1 and t_2 from the second projection memory, as shown in the Fig. 4 (1). In this situation, the active wordlines are shifted upwards by one step. That is, wl_1 and wl_2 are activated in the first projection memory but wl_0 and wl_1 are activated in the second projection memory. We use a control signal c_t to specify the relationship between the two sets of the activated wordlines of the two neighborhood projection memories and c_t can be determined by the input address.

Based on this observation, we propose two "smart" memory approaches which are named *decoder-mux* and *output-mux* respectively

Decoder-mux. In the first approach, called *decoder-mux*, we eliminate the decoder of the second memory and let it share the same decoder with the first memory by adding some configuration logic (which we also call decoder-mux) in between the two sets of memory wordlines. This logic configures the wordlines of the first projection memory (wla_i) to generate the wordlines for the next adjacent projection memory (wlb_i). The relationship between the wordlines of the two adjacent memories can be derived as

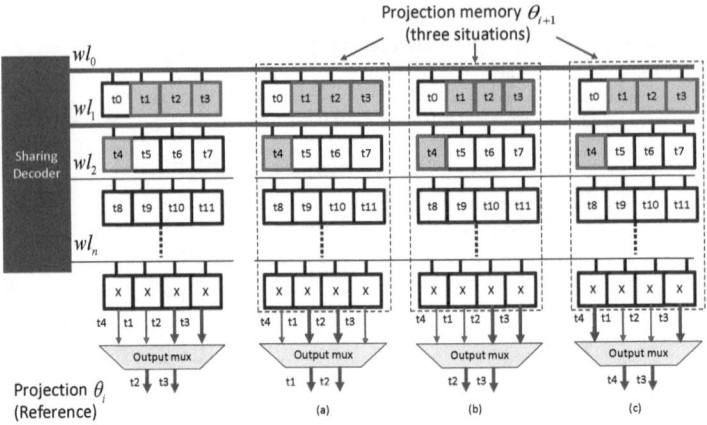

Fig. 6. Output-MUX. The memories are configured to output four pixels simultane-ously, and the *output mux* is used to select the required two pixels from the four outputs for the liner interpolation in each backprojection.

$$b_i = (-c_t) \cdot a_i + c_t \cdot a_{i+1}. \tag{20}$$

The configuration can be implemented using only AND and OR logic gates, which ensures the feasibility of the hardware implementation. In Fig. 5, we show an example of the configuration logic involving six wordlines. In this example, wla_1 and wla_2 are activated in the first memory array. After the decoder-mux block, either the same wordlines, wlb_1 and wlb_2, are activated in the second memory when $c_t = 0$ (Fig. 5 (a)), or the neighborhood wordlines, wlb_0 and wlb_1, are activated when $c_t = 1$ (Fig. 5 (b)).

Output-mux. In the alternative approach named *output-mux* the two memories still share the decoder but the configuration logic is located outside of the mem-ory (see Fig. 6). In this approach, memories are designed as the 1×4 consecutive access memories to output more elements than required. In this example, t_2, t_3 along with their nearest neighbors t_1 and t_4 are all read out from the memories. Then the configuration logic (*output-mux*) is used to select the appropriate two elements from the four outputs. In this approach, the active wordlines for the two memories are always the same in all the situations.

4.4   Horizontal and Vertical Parallel Backprojection

The method of *decoder-mux* and *output-mux* can be further extended to let more than two adjacent projection memories share one memory decoder. When more projection memories are involved, the address differences will be accumulated. As explained in the formulae (11), the address difference of the next k projec-tion memory from the first reference memory is increasing proportionally with k. Therefore, we will have to configure the smart memory design in order to

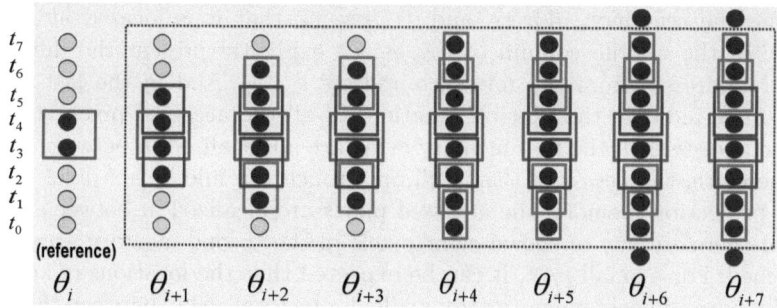

Fig. 7. Parallel Projection Memory Accessing. The highlighted two-pixel groups represent the beam pixels that have chances to be accessed in each projection memory.

accommodate the increased address differences if we want to let more than two adjacent projection memories share one memory decoder.

To exploit the proposed smart memory mechanisms to obtain superior hardware efficiency of the parallel backprojection, we propose two parallel approaches, that is, *horizontal and vertical parallel backprojection.*

Horizontal Parallel Backprojection. The horizontal parallel backprojection can perform more than two backprojections in parallel and all the involved projection memories share the same memory decoder using either *decoder-mux* or *output-mux* approach. Fig. 7 shows the example of accessing in eight adjacent projection memories. Assuming that the pixels addressed by the first memory addresses are t_3 and t_4, we highlight the possible locations of the two pixels accessed in the next seven memories. We observe that they are all clustered locally around t_3 and t_4, and are bounded by t_0 and t_7. For example, the pixels required for projection θ_{i+3} could be any two adjacent pixels within $[t_1, t_6]$. Required pixels spread out further from t_3 and t_4 for memories that are further away from the first memory as explained by formulae (11). Similar to the *output-mux* design shown in Fig. 6, we configure each projection memory as an 1×8 consecutive access memory to output all the shown eight pixels and use another 8-to-2 output-mux to select the appropriate two outputs from the eight outputs for each projection memory. In this way, all the eight memories could share the same decoder and seven memories decoders are saved. However, as the projection memories output more pixels than required, many memory outputs are actually wasted. An approach to use these wasted pixels is applying vertical parallel backprojection, as discussed next.

Vertical Parallel Backprojection. From (12) to (19), we discuss the address differences for performing the backprojections of four neighborhood pixels, that is, (x, y), $(x+1, y)$, $(x, y+1)$, $(x+1, y+1)$, concurrently. Backprojection of each pixel per projection angle requires one linear interpolation and involves memory accessing of two pixels, so totally it requires eight pixels to be accessed from each projection memory. To analyze the address distribution of these pixels, we compute all the involved addresses to the projection memories of projection angle θ_i and projection angle θ_{i+1} respectively, assuming $r/N = 1/4$. We let

t_{x,y,θ_i} be the reference address, and we assume that it is located at t_{13} (see Fig. 8). In the middle column of Fig. 8, we explicitly present the differences of other addresses from the reference address t_{x,y,θ_i}. And in the last column of Fig. 8 we indicate the possible locations of all the accessed pixels. It's seen that the addresses in the first memory array are all localized in between t_{11} and t_{15}, therefore, the access of them will only touch the middle six pixels. In the second projection memory, the accessed pixels are localized in between t_{20} and t_{26}, and therefore any of shown eight pixels in the second memory array could be touched. For a small r/N, it can be expected that the locations of accessing pixels in more adjacent memory arrays will also be localized in between the shown eight pixels. In this way, we support the vertical parallel backprojection which can perform the backprojections of multiple neighborhood pixels in parallel. The memory architecture needs no changes for the vertical parallel backprojection since we just take advantage of the unused memory outputs from the horizontal parallel backprojection. By implementing both horizontal and vertical parallel backprojection concurrently using the modified consecutive access memory, all the memory outputs are utilized and a much higher throughput is achieved.

5   Parallel Backprojection Architecture

The CT image reconstruction naturally lends itself to parallel processing since each backprojection can be processed independently. In this section, we will first introduce the conventional pipeline parallel backprojection architecture. Then we will develop a more advanced memory sharing pipeline parallel backprojection architecture based on the smart memory structures that we have introduced.

5.1   Parallel Pipeline BackProjection Architecture

An existing efficient architecture for projectionbased processing is the parallel pipeline backprojection engine (PPPE) [14, 1] due to its simplicity and potential speed. Fig. 9 (a) illustrates the structure of the PPPE based backprojection system, which employs an array of identical IPEs to reconstruct the image recursively, where each IPE performs the same tasks on a different projection. The input image is presented to each IPE on the pipelined image bus, one pixel at a time in a raster-scan format. In raster-scan format the x coordinate of the image is incremented every clock cycle and the y coordinate is incremented every line.

To start the operation, the first IPE in the pipeline is fed a blank image and adds the contribution of the first projection one pixel at a time. After the first IPE adds its contribution, it passes the pixel to the next IPE in the pipelined image bus and each IPE of the pipe adds its projection's contribution to the image. Therefore, each IPE_n in the pipe performs the backprojection for the angle θ_n, and add the resulting value to the input pixel, and then passes the pixel onto IPE_{n+1} as it receives another pixel from IPE_{n-1}. As the image pixel is sent through the pipelined array, the pixel value is reconstructed after accumulating the backprojected values from all the projections. The pipelined calculation and the raster-scan input allow high data throughput of one pixel per clock cycle.

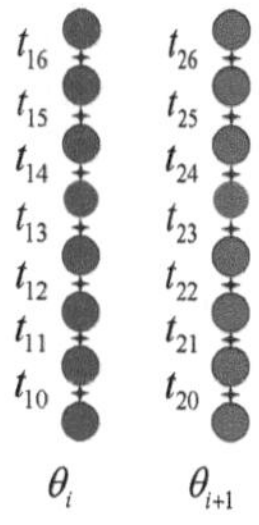

Beam Index	Index Difference to t_{x,y,θ_i}	Possibly Accessed Data
t_{x,y,θ_i}	0	t_{13}
t_{x+1,y,θ_i}	$\lvert\cos\theta_i\rvert \le 1$	$t_{12}\ t_{13}\ t_{14}$
$t_{x,y+1,\theta_i}$	$\lvert\sin\theta_i\rvert \le 1$	$t_{12}\ t_{13}\ t_{14}$
$t_{x+1,y+1,\theta_i}$	$\lvert\sin\theta_i + \cos\theta_i\rvert \le \sqrt{2}$	$t_{11}\ t_{12}\ t_{13}\ t_{14}\ t_{15}$
$t_{x,y,\theta_{i+1}}$	$\lvert\delta t_{\theta_i}\rvert \le 1.11$	$t_{21}\ t_{22}\ t_{23}\ t_{24}\ t_{25}$
$t_{x+1,y,\theta_{i+1}}$	$\lvert\cos\theta_i + \delta t_{\theta_i}\rvert \le 2.11$	$t_{20}\ t_{21}\ t_{22}\ t_{23}\ t_{24}\ t_{25}\ t_{26}$
$t_{x,y+1,\theta_{i+1}}$	$\lvert\sin\theta_i + \delta t_{\theta_i}\rvert \le 2.11$	$t_{20}\ t_{21}\ t_{22}\ t_{23}\ t_{24}\ t_{25}\ t_{26}$
$t_{x+1,y+1,\theta_{i+1}}$	$\lvert\sin\theta_i + \cos\theta_i + \delta t_{\theta_i}\rvert \le 2.52$	$t_{20}\ t_{21}\ t_{22}\ t_{23}\ t_{24}\ t_{25}\ t_{26}$

Fig. 8. Address Differences Analysis

5.2 Advanced Memory Sharing Parallel Pipeline Backprojection Architecture

If there are fewer IPE in the pipeline than angles (N_θ), then multiple passes through the IPE array are required to reconstruct the image. However, the performance will be decreased proportionally when the number of the IPE decreases. As an effective solution to increase the performance but minimize the hardware cost, we can modify the pipeline backprojection architecture to an more advanced memory-sharing based parallel pipeline backprojection engine (MSPPPE) by taking advantage of the our previous discussed horizonal and vertical backprojection methods. MSPPPE is also composed of a pipeline of identical image processing engines, however, each IPE will perform multiple backprojections to multiple pixels concurrently.

Base on the horizontal parallel backprojection, we let each IPE perform more than one backprojections simultaneously and each IPE needs to hold all of the involved projection data on-chip. So conventionally each projection memory is implemented as a multi-banking memory system in order to supply the data that are required in the parallel CT backprojection. Based on the above horizontal parallel backprojection approach, in each IPE we can combine all the projection data memory into one large memory block by locating them horizontally in parallel with each other so that all of these projection memories could share one memory decoder. In this way, the large overhead that associate with the multiple memory-banking design that were required in the parallel backprojection design can be eliminated. On the other hand, to take advantage of the vertical parallel backprojection, we increase the raster-scan bandwidth by letting more than one pixels pass through the pipeline simultaneously. Although the calculation of the contribution of every projection to every pixel needs to be performed in parallel, only the ALU needs to be duplicated to enable the parallel computing. The memory structure and its associate cost will be the same as above since we will just reuse the redundant output from the horizontal parallel backprojection.

The modified architecture is illustrated in Fig. 9 (b), where we show an example that the input image passes through the IPE on the pipelined image bus, four pixels at a time. Each IPE_n in the pipe performs eight adjacent backprojects from θ_i to θ_{i+7} to the current four pixels $(P(x,y), P(x+1,y), P(x,y+1), P(x+1,y+1))$,

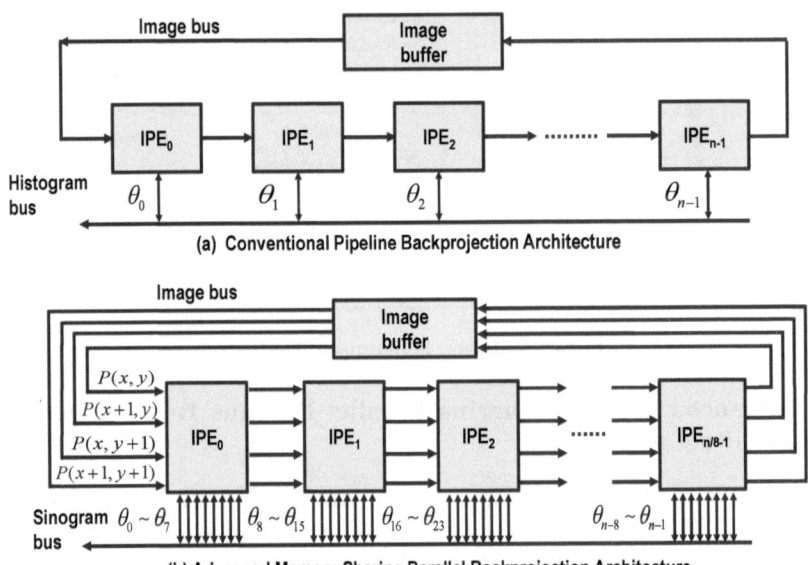

(a) Conventional Pipeline Backprojection Architecture

(b) Advanced Memory Sharing Parallel Backprojection Architecture

Fig. 9. Parallel Pipeline Backprojection Architecture

and then passes these pixels onto the IPE_{n+1} as it receives another four pixels from IPE_{n-1}. As these pixels are sent through the pipelined array, the pixel values are accumulated from the contributions of all the projections.

6   Design Automation

In this section we analyze the design space and describe our design automation framework for the hardware synthesis of a user-specified backprojection design point.

6.1   Design Tradeoff Space

Designing a CT image reconstruction system is a tradeoff problem involving algorithmic constraints, performance, hardware cost, and image accuracy. The discussion of address patterns in Section 3 shows that the ratio of image dimension size (r) and the projection numbers (N), r/N, is an important algorithm constraint. Smaller r/N indicates smaller adjacent address differences, which allows for more adjacent projection memories sharing the memory decoder, saving more hardware cost and computing latency. However, it also limits the use of the method in applications with larger image size r and/or fewer projection angles N. For larger r/N, the corresponding larger address difference will limit the number of projection memories that can share the decoder. For example, in Fig. 7, the last two projection memories of θ_{i+6} and θ_{i+7} may require to access two pixels at the two ends, which are not accessible along with other eight pixels from the 1×8 consecutive access memory. To solve this problem we could

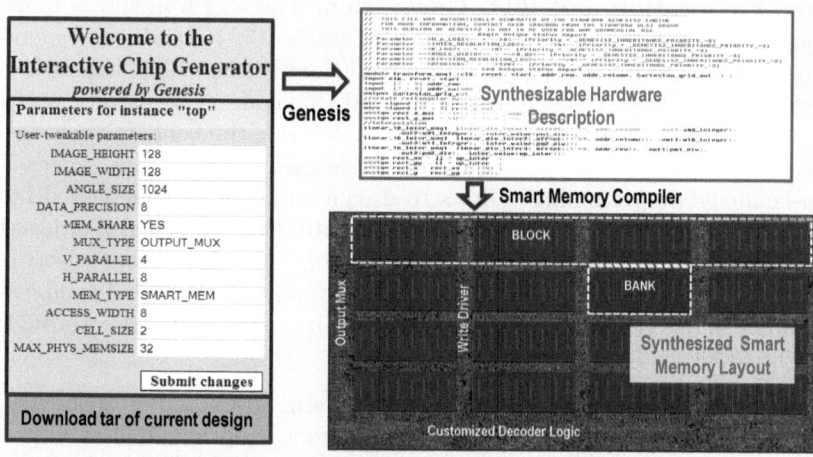

Gui Link: [http://genesis.web.ece.cmu.edu/gui/scratch/mydesign-13376.php]

Fig. 10. Smart Memory Design Framework

increase the memory access width and apply more complicated configuration logic. However, this would increase the hardware cost. Alternatively, to lower hardware cost we could assign the nearest neighborhood pixels if the requested pixels are not available, which would result in loss of image accuracy. This shows that different design decisions will result in different tradeoffs. The combination of these design choices constitutes a huge design space. Further, exploring the design tradeoff space requires customized memory designs, which are traditionally prohibitively expensive. Thus, a strong design automation tool is required to make the hardware synthesis feasible.

6.2   Chip Generator and Smart Memory Synthesizer

Application-specific LiM requires to tailor logic and memory design to application or algorithm specifics. Thus, a strong design automation tool is required to make the approach feasible, as hand-designing of LiMs is prohibitively expensive. We have developed a *design generation and design space exploration tool* which will automate the design of proposed customized smart memory blocks.

Our tool provides designers with a graphical user interface to select design parameters, and generate the corresponding hardware for the specified functionality. Un-specified parameters (free parameters) can be optimized by the system. A designer then evaluates the obtained designs and can explore the design space to optimize the design by varying the parameters. We encapsulate all of these design tradeoffs in our automatic design framework and build the backprojection smart memory synthesizer, the user interface is shown in Fig. 10. It enables an application designer to explore the design space to optimize the design by simply varying the parameters and automatically generates the optimized smart memory hardware IP.

Design Exploration and RTL Generation. The tool frontend is built using our chip generator infrastructure "GENESIS" [15, 16] and it's responsible for application interfacing, design optimization and efficient RTL generation. To achieve that, it allows designers to simultaneously code in two interleaved languages: a target language (SystemVerilog) to describe the behavior of hardware and a meta-language (Perl) to decide what hardware to use for given specs. This "dual-language programming" allows to design an entire parameterized family of LiM designs, all at once. Design parameters are set in graphical user interface (GUI) which is defined through XML files. An optimization engine selects optimized values for free parameters. The system supports hierarchical composition of modules and resolving of parameter constraints across modules through all hierarchy levels.

Smart Memory Compiler. The automated design framework discussed so far is capable of mapping LiM application specifications to optimal RTL. Our system also relies on a backend "smart memory" compiler to physically co-synthesize logic and memory. Today's embedded memory is typically synthesized using an SRAM compiler. But the use of commercial SRAM hardware IP is unable to incorporate application-specific customization that are required in the LiM design and also hinders comprehensive design space exploration. LiM physical synthesis is enabled by our *smart memory synthesis framework*, which is developed from the pattern construct based logic and memory co-design methodology [6, 7]. Using this framework, embedded logic in the LiM is synthesized together with the memory cells to a small set of pre-characterized layout pattern constructs. Lithographic compliance between the co-designed logic and memory ensures sub-20nm manufacturability of LiM circuits.

End-to-End LiM Design Framework. In our tool chain we are combining the architectural frontend and physical backend to build an end-to-end LiM design framework. Its input is the design specification and the output is ready to use hardware (RTL, GDS, .lib, .lef). When generating a specified design point, our framework also reports the area, power and latency and send them back to the frontend user interface, from which the designer can evaluate the resulting design and reset the design specs for redesign if necessary. Our LiM framework allows an application designer to generate the optimized "silicon" templates by simply tuning the "knobs".

7   Evaluation and Results

In this section, we evaluate the smart memory architectures with respect to area, power, latency, and accuracy. The design framework is used to generate various design points. Area and power are measured from the physical implementations of the design on a commercial sub-20nm SOI CMOS process at 500MHz and the shown results are all normalized.

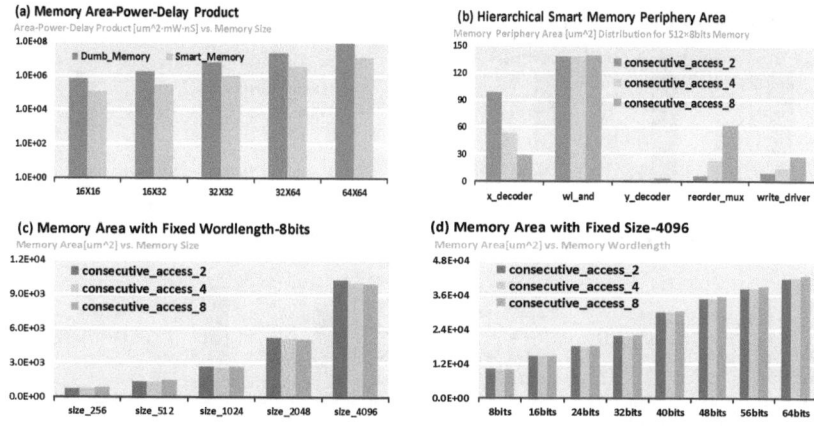

Fig. 11. Consecutive Access Memory Evaluation

7.1   Consecutive Access Memory Evaluation

The smart consecutive access memory is the basic memory structure that we use to implement various backprojection smart memory designs, therefore we evaluate its design efficiency first as shown in Fig. 11. To be consistent with the previous design, we implement the smart consecutive access memory to readout eight consecutive pixels from $1D$ data arrays from size 256 to size 4096. For comparison purpose, we also built the traditional multi-banking memory designs with the same functionalities. In Fig. 11 (a), we demonstrate the power-delay-product of the proposed smart consecutive access memory compared with the traditional multi-banking memory design (i.e., dumb memory), and it shows that the proposed smart memory are one order magnitude more efficient. To better understand the design structure of the smart consecutive access memory, we implement three different consecutive assess memories with different access bandwidths, that is, consecutive assess of two pixels, four pixels and eight pixels. We plot their hierarchical memory periphery area distribution in Fig. 11 (b). We see from the plot that the increase of the access width will decrease the area of the x-decoder while at the same time will increase the area of most other periphery circuit components (e.g., y-decoder, reorder-mux, write-driver and IO registers). This is because when the access width increases, the memory is getting wider and shorter as there will be more memory sub-banks sharing the x-decoder. Cell area is not plotted since it is assumed to be approximately the same for all the designs. For the same reason, the localized wordline AND logic (i.e., wl_and) area is the same for all the designs as each memory cell is associated with one AND gate in the customized x-decoder design. In Fig. 11 (c) and (d), we show the overall memory area for smart memory designs with different consecutive access widths at different memory sizes and different memory wordlengths respectively. One important observation is that the increase of the consecutive access width will not increase the overall smart memory area, and sometimes it even decreases the overall memory area for those larger-size memory designs (e.g., memory of

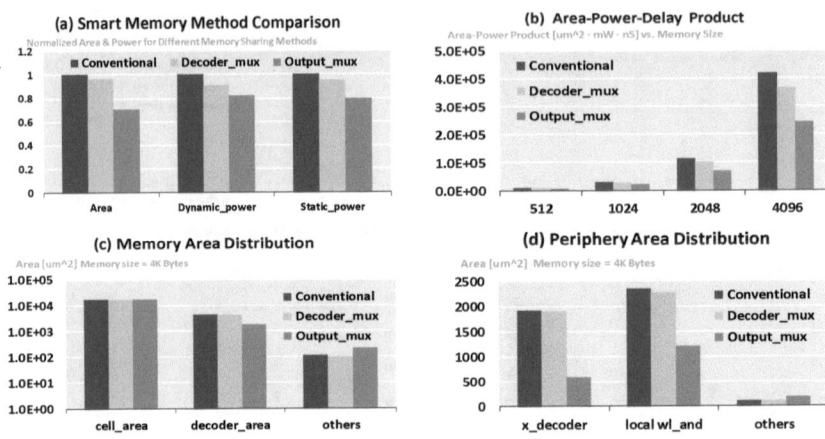

Fig. 12. Backprojection Smart Memory Evaluation

size 4096). This is because larger-size memory is associated with larger memory periphery circuits in the x-dimension (e.g., x-decoder) which can be reduced more in designs with larger access widths. However, the increase of the access width tends to cost more memory area for memories with larger wordlengh since in this situation the periphery circuits in the y-dimension (e.g., y-decoder) is getting larger and more complicated.

7.2   Backprojection Smart Memory Cost Evaluation

Decoder-mux and Output-mux Evaluation. In Fig. 12 (a), we first compare the hardware cost of two smart memory approaches (*decoder-mux* and *output-mux*) to the conventional rectangular access smart memory approach. The memories studied here have the size of 4,096-words and wordlength of 16 bits, and we only consider two memories implemented as 1×8 consecutive access memories sharing the decoder with each other. We observe that the *output-mux* approach is more cost-efficient as saves around 30% area and 20% power while *decoder-mux* only achieves around 5% area saving and 10% power saving. The similar results can be seen in Fig. 12 (b), in which we plot the overall area-power-delay of the three designs at four different memory sizes. As expected, *output-mux* approach saves on average $20\% - 40\%$ in terms of area-power-delay product. On the other hand, *decoder-mux* performs much worse compared with the *output-mux*. The reason is that in *decoder-mux* each wordline is accompanied by a set of configuration logic (two AND gates and one OR gate), and each set of logic communicates with its local wordline. This explains also why *decoder-mux* achieves relatively higher power-efficiency compared to its area-efficiency. In contrast, *output-mux* only requires a single large configuration logic at the memory output while its memories have large access width as they output more pixels than required. Due to the superiority of the *output-mux* method, it will be used for our backprojection system in the following discussions.

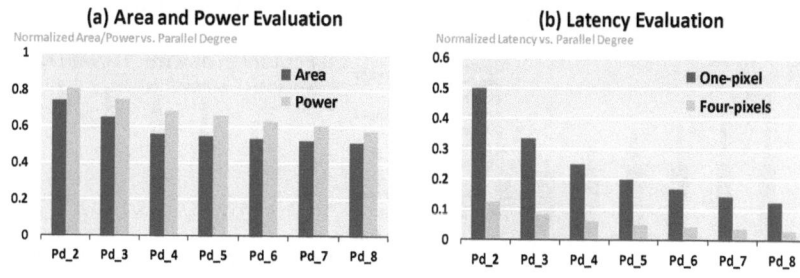

Fig. 13. Memory Sharing Parallel Pipeline Architecture Evaluation

As the main idea of the memory sharing strategy is to reduce the hardware cost by sharing the x-decoder, in order to understand the distribution of the hardware cost of the different components in the memory structure, in Fig. 12 (c) we plot the hierarchical memory area for all the three methods. It is observed that although memory cell array occupies most of the memory area, the periphery area also accounts for a large proportion of overall memory area. As the memory cell area of the three designs are the same, in Fig. 12 (d) we particularly plot the hierarchical memory periphery area for the three methods and we see that the memory periphery is dominated by the x-decoder and the embedded localized wordline AND logic (i.e., wl_and) gates. As we discussed in 4.2, the localized wordline AND logic (i.e., wl_and) gates are tightly integrated with the memory cell for local wordline activation. As can be seen, both of the decoder area and the local wl_and gates area are largely reduced in the output-mux approach as they can be directly shared by all the memory banks.

Parallel Backprojection Architecture Evaluation. In Fig. 13 (a) we evaluate the hardware cost of the MEPPPE memory architecture for reconstructing a 256×256-size image from 1,024 projections. The x-axis is the parallel degree P_d, which is defined as the number of adjacent backprojections that are performed in each IPE concurrently and its value varies from two to eight. In our implementation these P_d projection memories will all share the same memory decoder. The y-axis shows the relative area and power compared to the conventional design where no memory sharing strategies are used. We see that more than 40% area savings and more than 30% power savings can be achieved with the increase of P_d. Fig. 13 (b) shows that the latencies are decreasing proportionally with the increase of P_d as expected. Moreover, we achieve a four times performance improvement by computing four pixels in parallel in each IPE.

7.3   Backprojection Accuracy Evaluation

As we gain in both of hardware cost and performance, the impact on accuracy needs to be evaluated. In Fig. 14 (a), we show the distribution of the locations of the accessed data in eight adjacent projection memories for a real application. We first observe that the locations of the accessed data in eight adjacent projection

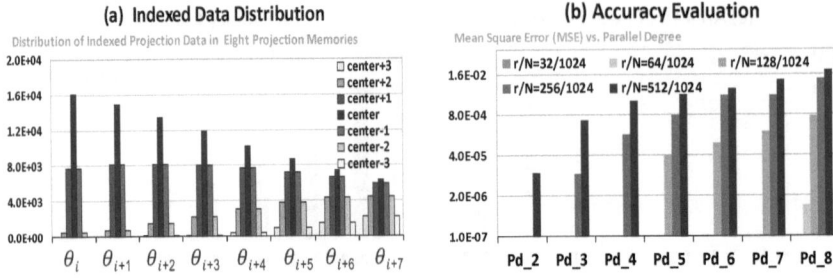

Fig. 14. Image Accuracy Evaluation

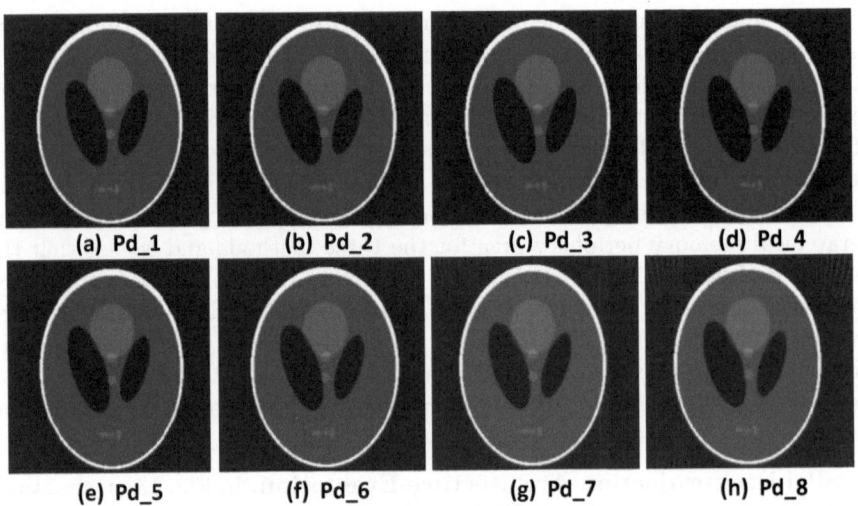

Fig. 15. Display of Reconstructed Image

memories are all localized to the location of *center*. Therefore we could design all the eight projection memories to output the pixels within the range between *center* − 3 and *center* + 3 so that they could share one memory decoder based on our *output-mux* design. However, it can also be seen that the range of possible locations of the accessed data are increasing when we go from projection memory of angle i to the projection memory of angle $i + 7$. For example, starting from projection angle θ_{i+4}, all the shown seven locations will be intensively touched. It can be expected that if we let more adjacent projection memories share the decoder, they could require pixels that are beyond the smart memory outputs. We could approximately assign the nearest pixels if the required pixels are not available but it will then sacrifice the resulting image quality in such situations.

We measure the mean square error (MSE) of the reconstructed image compared to the reference image and plot the results in Fig. 14 (b) for parallel degrees (P_d) from one to eight. As expected, the error increases when either P_d or algorithm parameter (r/N) increases. This is because that we let P_d

projection memories share the same memory decoder, and it will introduce error if the address differences of these P_d projection memories are not small enough which could happen when P_d and (r/N) are large. In our implementation, we carefully manipulate the data precision so that the numerical errors can be ignored in the accuracy comparison. In Fig. 15 we display the reconstructed head phantom images from hardware simulation, which indicates fairly high image quality for all the studied parallel degrees. We also observe the gradual deterioration of the image quality for higher parallel degree, which allows us to tradeoff image accuracy with hardware cost in applications where minor distortion is acceptable.

8   Conclusion

The emergence of construct-based design facilitates the robust synthesis of cost-effective smart memory blocks that are customized for specific applications. This cutting-edge design methodology creates opportunities to re-design algorithms and re-architect the hardware structure to match the advanced technology capabilities. In this paper we propose smart memory architectures and the end-to-end design framework to implement them for the CT image reconstruction problems. The results in a sub-20nm CMOS process demonstrate significant improvements in area, power and performance. Moreover, we present the opportunities to tradeoff hardware cost with acceptable image accuracy based on appropriate algorithm tuning. This paper demonstrates that the embedded memories in data-intensive computing can exploit the smart memory design methodology and the inherent address pattern of the algorithm to achieve superior power and performance efficiency.

Acknowledgement. The authors acknowledge the support of the C2S2 Focus Center, one of six research centers funded under the Focus Center Research Program (FCRP), a Semiconductor Research Corporation entity.

References

1. Agi, I., Hurst, P.J., Current, K.W.: An Image Processing IC for Backprojection and Spatial Histogramming in a Pipelined Array. IEEE Journal of Solid-State Circuits 28(3), 210–221 (1993)
2. Srdjan, C., Miriam, L., Miller, E., Trepanier, M.: Parallel-Beam Backprojection: An FPGA Implementation Optimized for Medical Imaging. FPGA (2002)
3. Chen, C., Cho, Z., Wang, C.: A Fast Implementation of the Incremental Backprojection Algorithms for Parallel Beam Geometries. IEEE Transactions on Nuclear Science 43(6), 3328–3334 (1996)
4. Zhu, Q., Turnerz, E.L., Bergery, C.R., Pileggi, L., Franchetti, F.: Application-Specific Logic-in-Memory for Polar Format Synthetic Aperture Radar. In: IEEE Conference on High Performance Extreme Computing, HPEC (2011)

5. Zhu, Q., Bergery, C.R., Turnerz, E.L., Pileggi, L., Franchetti, F.: Polar Format Synthetic Aperture Radar in Energy Efficient Application-Specific Logic-in-Memory. In: IEEE International Conference on Acoustics, Speech and Signal Processing (ICASSP), pp. 1557–1560 (2012)

6. Morris, D., Rovner, V., Pileggi, L., Strojwas, A., Vaidyanathan, K.: Enabling Application-Specific Integrated Circuits on Limited Pattern Constructs. In: Symp. VLSI Technology (2010)

7. Morris, D., Vaidyanathan, K., Lafferty, N., Lai, K., Liebmann, L., Pileggi, L.: Design of embedded memory and logic based on pattern constructs. In: Symp. VLSI Technology (2011)

8. Luiz, M.C.B., Felipe, M.G.F., Vladimir, C.A., Claudio, L.A.: Reconfigurable Hardware for Tomographic Processing. In: Proceedings of the XI Brazilian Symposium on Integrated Circuit Design, pp. 19–24 (1998)

9. Jang, B., Kaeli, D., Do, S., Pien, H.: Multi GPU Implementation of Iterative Tomographic Reconstruction Algorithm. In: International Symposium on Biomedical Imaging (ISBI), pp. 185–188 (2009)

10. Yu, H.Q.: Memory Architecture for Data Intensive Image Processing Algorithms in Reconfigurable Hardware. Master Thesis (2003)

11. Cho, Z.H., Chen, C.M., Lee, S.Y.: Incremental Algorithm - A New Fast Backprojection Scheme for Parallel Beam Geometries. IEEE Transactions on Medical Image 9(2), 207–217 (1990)

12. Zhu, Q.L., Vaidyanathan, K., Shachamy, O., Horowitz, M., Pileggi, L., Franchetti, F.: Design Automation Framework for Application-Specific Logic-in-Memory Blocks. In: Application-Specific Systems, Architectures and Processors (ASAP), pp. 125–132 (2012)

13. Murachi, Y., Kamino, T., Miyakoshi, J., Kawaguchi, H., Yoshimoto, M.: A Power-Efficient SRAM Core Architecture with Segmentation-Free and Rectangular Accessibility for Super-Parallel Video Processing. In: IEEE International Symposium on VLSI Design, Automation and Test (VLSI-DAT), pp. 63–66 (2008)

14. Hinkle, E.B., Sanz, J.L.C., Jain, A.K., Petkovic, D.: P3E: New life for projection-based image processing. Journal of Parallel and Distributed Computing 4(1), 45–78 (1987)

15. Shacham, O.: Chip multiprocessor generator: automatic generation of custom and heterogeneous compute platforms. PhD Thesis, Stanford (2011)

16. Stanford genesis2 web site, http://genesis2.stanford.edu/mediawiki/index.php

Trinocular Stereo Vision Using a Multi Level Hierarchical Classification Structure[*]

Andy Motten[1], Luc Claesen[1], and Yun Pan[2]

[1] Expertise Centre for Digital Media, Hasselt University – tUL – iMinds
Wetenschapspark 2, 3590 Diepenbeek, Belgium
`firstname.lastname@uhasselt.be`
[2] Institute of VLSI Design, Zhejiang University
Hangzhou, China
`panyun@vlsi.zju.edu.cn`

Abstract. A real-time trinocular stereo vision processor is proposed which combines a window matching architecture with a classification architecture. A pair wise segmented window matching for both the center-right and center-left image pairs as their scaled down image pairs is performed. The resulting cost functions are combined which results into nine different cost curves. A multi level hierarchical classifier is used to select the most promising disparity value. The classifier makes use of features provided by the calculated cost curves and the pixels' spatial neighborhood information. Evaluation and classifier training has been performed using an indoor dataset. The system is prototyped on an FPGA board equipped with three CMOS cameras. Special care has been taken to reduce the latency and the memory footprint.

Keywords: trinocular stereo camera, real-time matching, confidence metric, computer vision, system-on-chip, FPGA, SoC.

1 Introduction

Trinocular vision makes use of three cameras to calculate a disparity space image (DSI). The DSI is generated by pairwise matching the images from the different cameras which is based on a local window based stereo matching architecture.

An improvement of occlusion handling in trinocular vision compared to stereo vision is achieved by Mozerov [1]. The main idea is based on the assumption that any occluded region in a matched stereo pair (center-left images) in general is not occluded in the opposite matched pair (center-right images). They use a global optimization technique to derive the composite DSI. Bidirectional matching using trinocular stereo is used by Ueshiba [2] to detect half-occlusions and to discard false matches. It uses a cumulative cost function derived from a summation of both cost curves.

[*] This research has been sponsored in part by the BOF (Bijzonder Onderzoeks Fonds uHasselt), Flanders FWO (Fonds voor Wetenschappelijk Onderzoek) and Chinese MOST (Ministry of Science and Technology) project number G.A.063.10.

A. Burg et al. (Eds.): VLSI-SoC 2012, IFIP AICT 418, pp. 45–63, 2013.

The method presented in this paper likewise calculates several DSI's. However, instead of combining them, a hierarchical classifier is used to select the most likely disparity for each pixel in the final DSI. The matching algorithm is based on the adaptive-weight algorithm proposed by Yoon [3], which adjusts the support weight of each pixel in a fixed sized window. The support weights are depending on the color and the spatial difference between each pixel in the window and the center pixel. Dissimilarities are computed based on the support weights and the plain similarity scores. Their experiment indicates that a local based stereo matching algorithm can produce depth maps similar to global algorithms. A hardware implementation using the same ideas is published by Motten in [4].

For each matching result, a confidence metric is calculated. A good comparison between different confidence metrics can be found in the evaluation paper of Hu [5]. Confidence metrics suitable for hardware implementation can be found in [6]. They conclude that neighboring pixels contain valuable information to distinguish good matches from bad ones.

Recently many stereo implementations have been proposed for hardware implementations. A real-time FPGA-based stereo vision system is presented by Jin [7] that makes use of the census transform. Their system includes all the pre- and post-processing functions such as: rectification, LR-check and uniqueness test in a single FPGA. Another extensive implementation can be found in [8]. They divide the problem into two parts: first a rough depth map is constructed using a segmentation based SAD window comparison, second a disparity refinement module identifies false matches and replaces them with new estimates. Hardware implementations of a trinocular disparity processor are limited. An implementation using the summation of SAD's from both image pairs can be found in [9].

This paper combines the strengths of an advanced stereo vision system with a two-scale adaptive window SAD incorporated in a trinocular setup.

2   System Overview

2.1   General Architecture

The trinocular disparity processor takes three images that have been taken by three cameras that have a vertical alignment and a horizontal offset (see Fig. 1). The objective is to calculate a disparity space image (DSI) where dark pixels represent a distance further away from the cameras and a light pixel represents a distance closer to the camera.

Objects will appear on the same horizontal line (the epipolar line) on all images. The horizontal distance between the same objects on the center image and the left (or right) image is called the disparity. If calibrated correctly, the disparity of an object between the center-left and the center-right image pair is the same. This characteristic can be used to discard false matches using bidirectional matching [2] or to improve the quality of the disparity space image (DSI) especially in occluded regions [1].

The architecture consists of three main blocks (see Fig. 2). The first block captures the pixel streams, generates the scaled images and places them in multiple parallel on-chip memories. The second block performs a pair wise window comparison of the different streams using a binary adaptable SAD cost aggregation [8]. The third block calculates its confidence for each data stream and selects the final disparity value.

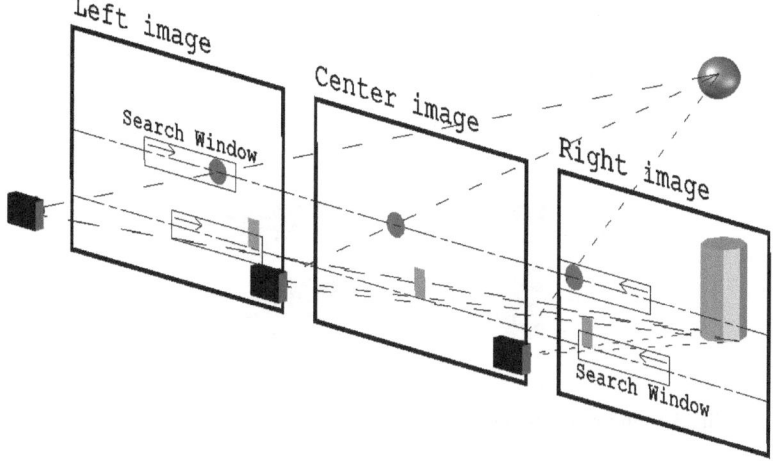

Fig. 1. Trinocular disparity processor setup

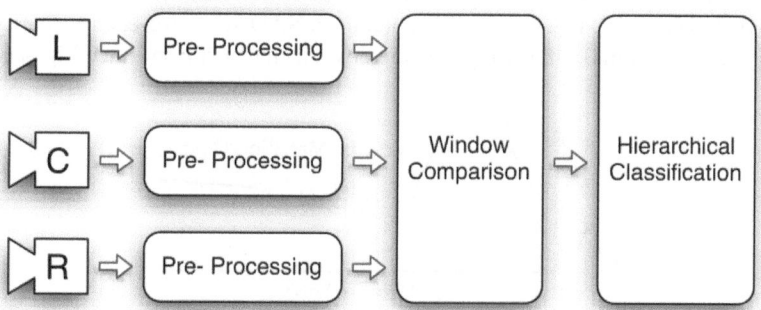

Fig. 2. Global architecture

On several places, this architecture makes use of a binary support window. When using a fixed window shape, depth continuity is implicitly assumed across this window. This assumption is false at depth edges where parts of the window belong to different depth levels. A more conservative assumption is to only assume depth continuity across pixels with similar color. Yoon [3] proposed an adaptive weight algorithm which gives a support weight to each pixel in a window. In order to save system on chip resources, an alternative has been proposed where the support weights are chosen as binary values [4]. A value of '0' means that this pixel doesn't belong to the support window of the center pixel and '1' means that this pixel belongs to the support window of the center pixel. This is called the binary support window (1). It is calculated by taken the absolute difference of the chroma color components (C_B, C_R) of all pixels q belonging to the rectangular window centered in pixel p.

$$w = \begin{cases} 0 \; if(\; |C_B(q) - C_B(p)| + |C_R(q) - C_R(p)|) > threshold \\ 1 \; otherwise \end{cases} \tag{1}$$

This binary support window is used when comparing different windows with each other (2). Instead of comparing a complete window, only the pixels where the support window is '1' (white) will be taken into account (see Fig.3).

$$SAD = \sum_{i=1}^{window\ size} w * absolute\ difference(i) \tag{2}$$

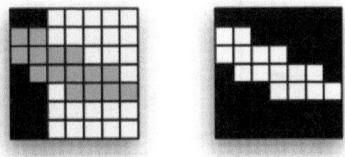

Fig. 3. Window content (left) and resulting binary support window (right)

Each window of the center image needs to be matched with multiple windows of the left or the right camera. For every window that needs to be matched, a SAD calculation is performed. The larger the disparity search width, the more SAD calculations are needed. The result is an array that contains a SAD score for each disparity value (usually starting from 0), this array is also known as the cost curve (see Fig. 4).

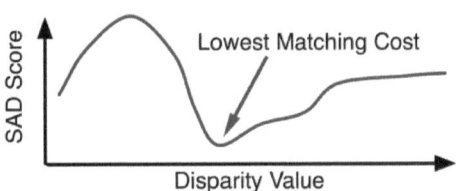

Fig. 4. Cost curve example

In this paper, C1 stands for the lowest SAD score (the minima of the Cost Curve). C2 stands for the second lowest SAD score, and so on. Their corresponding depths are indicated by D1 and D2. Most matching algorithms calculate the disparity from the cost curve using a "Winner Takes All" (WTA) approach. Doing so, the minima of the cost curve (C1) will become the calculated disparity D1.

In this architecture nine different cost curves are calculated for each pixel in the DSI. The first step is to calculate the cost curve the center-right (SAD_{CR1}) and center-left (SAD_{CL1}) image pairs as their scaled down image pairs (SAD_{CR0} and SAD_{CL0}). The second step is to calculate the summation of these cost curves (3).

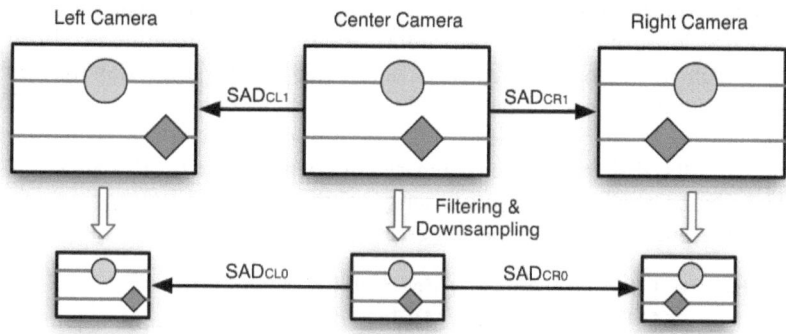

Fig. 5. Different window matching

$$\begin{cases} SAD_{CL1}, SAD_{CR1}, SAD_{CL0}, SAD_{CR0} \\ SAD_{CLR1} = SAD_{CL1} + SAD_{CR1} \\ SAD_{CLR0} = SAD_{CL0} + SAD_{CR0} \\ SAD_{CL01} = SAD_{CL1} + SAD_{CL0} \\ SAD_{CR01} = SAD_{CR1} + SAD_{CR0} \\ SAD_{CLR01} = SAD_{CLR1} + SAD_{CLR0} \end{cases} \tag{3}$$

3   Hierarchical Classification

In the previous section it is explained that nine disparity values are generated for each pixel (3). In order to select one of them for generating the DSI, a two level hierarchical classifier is constructed (see Fig. 6). In the first level of the hierarchy, the disparity values are investigated independently of each other. For each disparity value a binary confidence classifier is constructed using the methods presented in [6]. These confidences are passed on to the second level classifier which selects the disparity to use, or indicates that no disparity has been found.

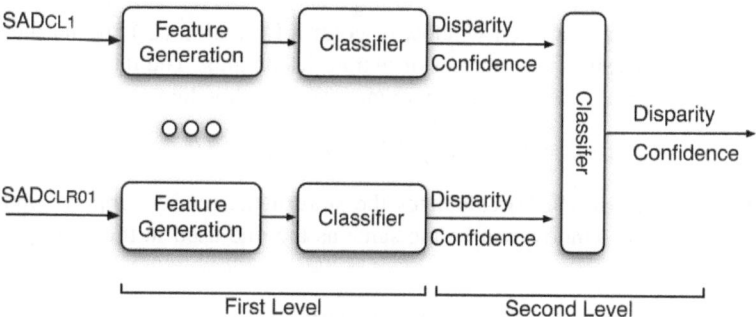

Fig. 6. Hierarchical classification

For each level of the hierarchy, a different set of features is needed for classification. The first level of classifiers uses information obtained from the pixel neighborhood and from its corresponding cost curve. The second level classifier uses the generated binary confidence values together with the agreement between the different disparity values. A binary confidence value of '1' indicates a strong confidence in the correctness of the disparity value. A binary confidence value of '0' indicates a weak confidence in the correctness of the disparity value.

Three different datasets have been used to verify the results:

- Tsukuba   [10]:   384 x 288 (Maximal disparity of 30).
- Teddy   [11]:   450 x 375 (Maximal disparity of 30).
- Art   [12]:   695 x 555 (Maximal disparity of 30).

In order to train a classifier, it is needed to define a target output. In this case, the preferable output would be a Boolean value indicating the correctness of the disparity value (the confidence value). A pixel is defined to be correctly matched with its corresponding disparity when the calculated disparity (D_c) and the real disparity (D_r) do not differ more than one unit disparity value (4, 5).

$$Error\ map(i) = \begin{cases} 1\ if\ D_c\ (i) \notin [D_r(i), D_r(i) \pm 1] \\ 0\ otherwise \end{cases} \tag{4}$$

$$Error_{Th1} = \sum_{i=1}^{image\ size} Error\ map(i) \tag{5}$$

3.1   Feature Generation

The features for the first level of classification are proposed in [6]. Their objective lies in accommodating the classification of the disparity stream for the first level of classification.

The matching cost (MC) is the minimum value of the cost curve. A high score will be a good indication of a wrong depth value.

$$MC = C_1 \tag{6}$$

The texture (TEX) uses a fixed window of color information (C_i) around the investigated pixel and measures the amount of texture it contains. The intuition behind it is that textureless regions will provide more incorrect depth values.

$$TEX = max_{window}(C_i) - min_{window}(C_i) \tag{7}$$

The segmentation size (SEG) calculates the sum of the binary support window (1). This binary support window can be the same as the one used in the cost aggregation phase.

$$SEG = \sum_{i=1}^{window\ size} w \tag{8}$$

The following two features make use of neighborhood information of the disparity space image (DSI). In order to calculate them, a buffer is needed to store several lines of the DSI. The size of this window depends on the size of the neighborhood and the width of the image.

The sum of neighboring depths differences (SNDD) uses a fixed window of depths around the investigated pixel and calculates the depth differences in this window.

$$SNDD = \sum_{i=1}^{window\ size} |D_1(i) - D_1(center)| \tag{9}$$

The sum of neighboring depths differences binary window (SNDDBW) is similar to SNDD, but instead of using a fixed window it uses only the neighboring pixels, which have a similar color. This is a different usage of the binary support window (1).

$$SNDDBW = \frac{\sum_{i=1}^{window\ size} w * |D_1(i) - D_1(center)|}{\sum_{i=1}^{window\ size} w} \tag{10}$$

The following features are designed for multi stream classification. They take the confidence value generated from the first level of classification and provide a feature which objective lies in accommodating the selection of the best disparity stream.

The sum of streaming depths differences (SSDD) calculates the depth difference between the different disparity streams taking the confidence value into account.

$$SSDD = \sum_{i=1}^{streams} confidence_i * |D_1 - D_1(i)| \tag{11}$$

The sum of streaming confidences (SSC) calculates the number of streams which have a positive confidence.

$$SSC = \sum_{i=1}^{streams} confidence_i \tag{12}$$

3.2   Classification Methods

The first level classifier consists of a decision tree (DT) for each disparity stream individually. The decision tree is a top-down tree structure consisting of internal nodes, leaf nodes, and branches. Each internal node represents a decision on a feature, and each outgoing branch corresponds to a possible outcome. Each leaf node represents a class (0 or 1 in this case). The main advantage of a decision tree is the ease of interpretation and implementation, while still being able to separate hard to separate classes. In the example of Fig. 7, two classes are separated by the class boundary which is constructed using a small DT.

The second level classifier chooses the final disparity from the different disparity streams by choosing the one with the lowest SSDD. A decision tree classifier is trained to construct a confidence value for the final disparity value.

Both classification methods are easily implemented in hardware without using many resources.

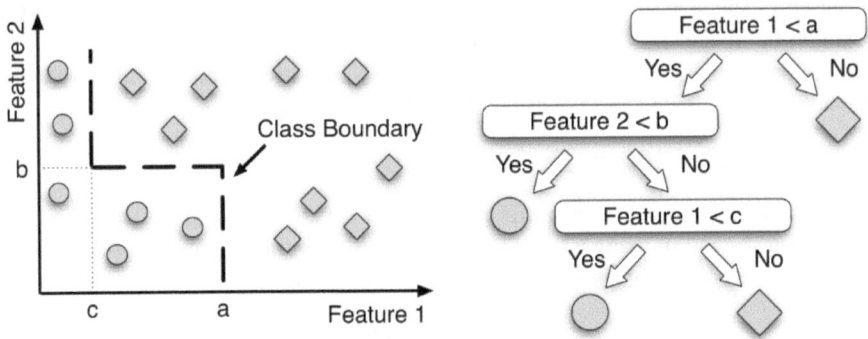

Fig. 7. Decision tree example: two-dimensional feature space (left) and resulting DT (right)

3.3   First Level Classification Evaluation

Matlab has been used to generate and classify the different features using five-fold cross validation. The dataset is split into five sets, where four sets are used for training the classifier and one set is used for validating the results. This is repeated five times, such that each of the five sets is used exactly once as validation set. The five validation results are averaged to generate the final result.

For each dataset, cost curves have been calculated using the SAD aggregation method for a binary adaptable window with four different selection thresholds and a window size of 7x7.

The features are indicated in the following table by their acronym and by the subscript of their parameters: e.g. $SNDDBW_{21,8}$ means SNDDBW with a window size of 21x21 and a Chroma threshold of 8.

The results of the first level classification can be seen in table 1. For every test, the feature is shown which is most important in the construction of the tree. From this table, we can see that the error rate between the different disparity streams is different; by summation of the SAD's a more correct DSI is constructed.

From Fig. 8 we can see that the different DSI's have different areas which are indicated as correct. The center-left image comparison (CL1) provides a good result across borders where the area on the left side of the border is further away then the area on the right side of the border. The center-right image (CR1) provides good results when the border is reversed. A summation of both SAD's (CLR1) gives a lower global error rate, but the borders are less clear.

As expected, the scaled down image pair comparisons (CL0 – CR0 – CLR0) provides better results on large texture less areas compared with the normal size image pair comparisons (CL1 – CR1 – CLR1). This is particular true for the background. However they have problems finding the correct disparity value for small objects, like the bars on the lamp. The summation of the SAD's generated by the normal size and the scaled downs image pairs (CL01 – CR01) gives a lower global error rate, keeps small details and has a better result with texture less areas.

Table 1. First level classification

Data Stream	Data	Feature Name	Error Rate DSI (Th1)	Misclassification Binary Classifier
CL0	Tsukuba	$SNDDBW_{21,8}$	27.30%	17.13%
	Teddy	$SNDDBW_{21,8}$	13.35%	5.35%
	Art	$SNDDBW_{21,8}$	19.65%	10.60%
CR0	Tsukuba	$SNDDBW_{21,8}$	29.43%	21.15%
	Teddy	$SNDDBW_{21,8}$	18.26%	6.12%
	Art	$SNDDBW_{21,8}$	19.27%	10.64%
CL1	Tsukuba	$SNDDBW_{21,8}$	22.98%	11.51%
	Teddy	$SNDDBW_{21,8}$	14.53%	13.39%
	Art	$SNDDBW_{21,8}$	23.74%	18.12%
CR1	Tsukuba	$SNDDBW_{21,8}$	25.05%	11.88%
	Teddy	$SNDDBW_{21,8}$	23.55%	16.25%
	Art	$SNDDBW_{21,8}$	24.45%	19.53%
CL01	Tsukuba	$SNDDBW_{21,8}$	22.92%	14.77%
	Teddy	$SNDDBW_{21,8}$	11.60%	6.05%
	Art	$SNDDBW_{21,2}$	19.11%	10.21%
CR01	Tsukuba	$SNDDBW_{21,8}$	24.83%	18.53%
	Teddy	$SNDDBW_{21,8}$	17.06%	6.83%
	Art	$SNDDBW_{21,2}$	18.99%	11.50%
CLR0	Tsukuba	$SNDDBW_{21,8}$	25.19%	20.06%
	Teddy	$SNDDBW_{21,8}$	14.02%	7.47%
	Art	$SNDDBW_{21,8}$	16.30%	10.13%
CLR1	Tsukuba	$SNDDBW_{21,8}$	22.40%	11.85%
	Teddy	$SNDDBW_{21,8}$	18.15%	21.59%
	Art	$SNDDBW_{21,8}$	19.12%	14.06%
CLR01	Tsukuba	$SNDDBW_{21,2}$	22.52%	16.67%
	Teddy	$SNDDBW_{21,8}$	14.17%	8.10%
	Art	$SNDDBW_{21,8}$	16.73%	10.45%

This indicates that by combining the DSI's, we could obtain a higher quality DSI. However before combining them, we need to know which part of each individual DSI is correct. A binary classifier is constructed to provide a confidence value for each DSI. The more correct this classifier, the more success we will have with the combined DSI. Depending on the dataset, a misclassification rate between 5% and 20% is obtained. This could be improved by using a more discriminative classifier like an artificial neural network [6].

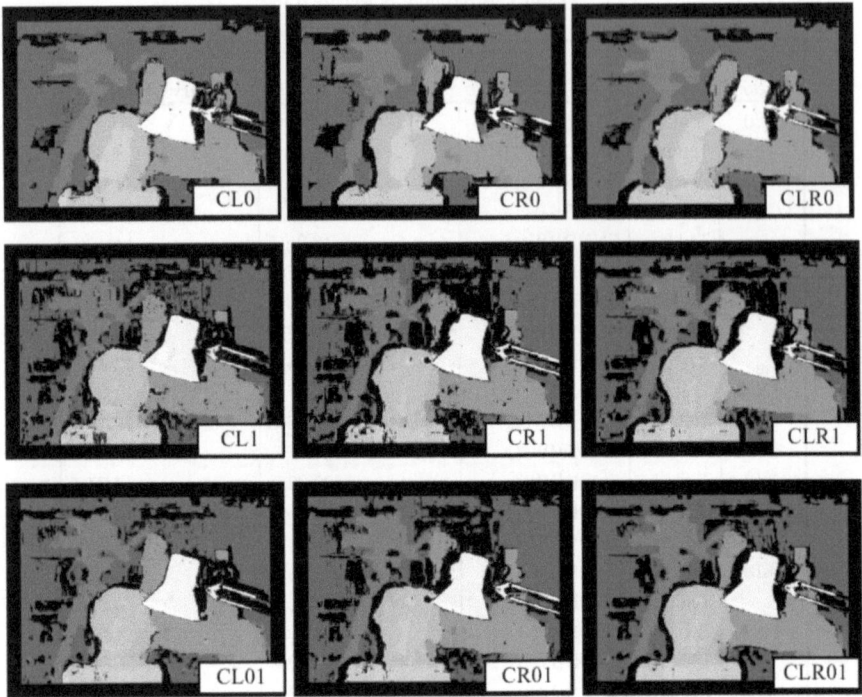

Fig. 8. Depth map quality of the Tsukuba dataset for a fixed window size of 7x7 after the first level of classification (black pixels indicate a confidence value of zero)

3.4   Second Level Classification Evaluation

The goal of this classification level is to select the most promising disparity value (13). The input of this classification level is the SSDD for each disparity stream. The output of this classification level is a disparity selection. A confidence value is afterwards generated using the same method as for each individual stream although using SSC as an extra input feature.

$$Disparity = \min_{i:1 \rightarrow streams} SSDD\,(i) \qquad (13)$$

An exhaustive search is performed in order to know which combination of streams provides the highest disparity improvement. A selected set of results can be seen in table 2. The results indicate that, by combining extra streams, the classification rate for all investigated datasets are improved. From Fig. 9 we can see that the addition of extra streams improves the quality of the DSI. The trinocular setup improves the DSI most noticeably at occluded regions. The scaled image improves the disparity map at parts with little texture.

Table 2. Second level classification

Data Stream	Data	Error Rate DSI (Th1)	Misclassification Binary Classifier
CL1 - CR1 - ...	Tsukuba	16.50%	12.90%
CL01 - CR01 - ...	Teddy	7.82%	7.18%
CLR1 - CLR0	Art	11.86%	10.09%
CL1 - CR1 -...	Tsukuba	16.52%	12.47%
CL0 - CR0	Teddy	7.74%	6.71%
	Art	12.43%	11.50%
CL1 - CR1 - ...	Tsukuba	16.85%	12.57%
CL01 - CL01	Teddy	8.05%	7.23%
	Art	13.09%	11.38%
CL01 - CR01 - ...	Tsukuba	17.25%	13.22%
CLR1 - CLR0	Teddy	7.58%	6.47%
	Art	11.82%	10.99%

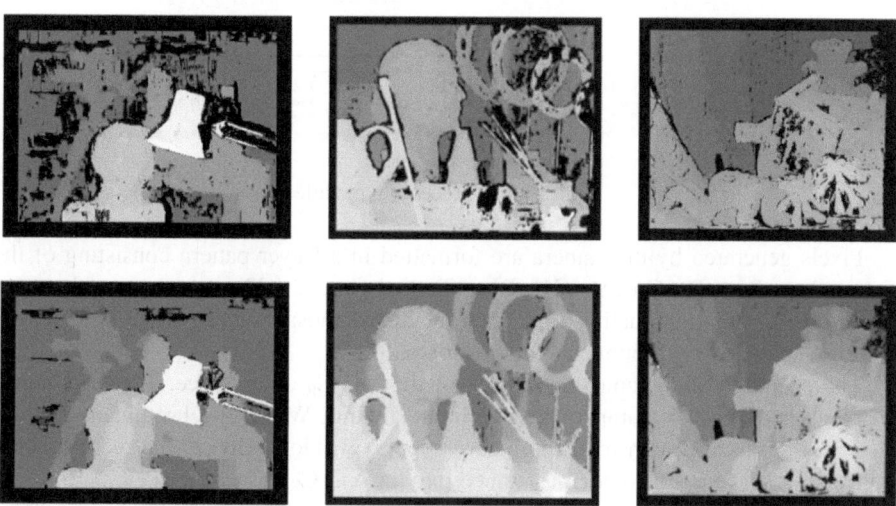

Fig. 9. Depth map quality of the investigated datasets (Tsukuba, Teddy, Art). Comparison of DSI generated from CL0 data stream (Top row) and DSI generated from the combination of CL0, CL1, CR0 and CR1 data streams (Bottom row).

4   System Design

The hardware architecture consists of three main modules. First a filter and sub sampling module has been added to the pre-processing module [8] so that a scaled image is generated with one-fourth the size of the original image. Second the window matching module is modified from [8] to allow for multiple data stream matching. Third a hierarchic classification module is constructed to select the most promising disparity from the different disparity results.

4.1   Pre-Processing Module

The pre-processing module (see Fig. 10) consists of four different entities for each pixel stream: first a Bayer demosaicing algorithm is used to reconstruct the color image, next a rectification module is used to remove lens distortion and perform trinocular calibration and lastly the image is filtered and down sampled to generate a scaled image.

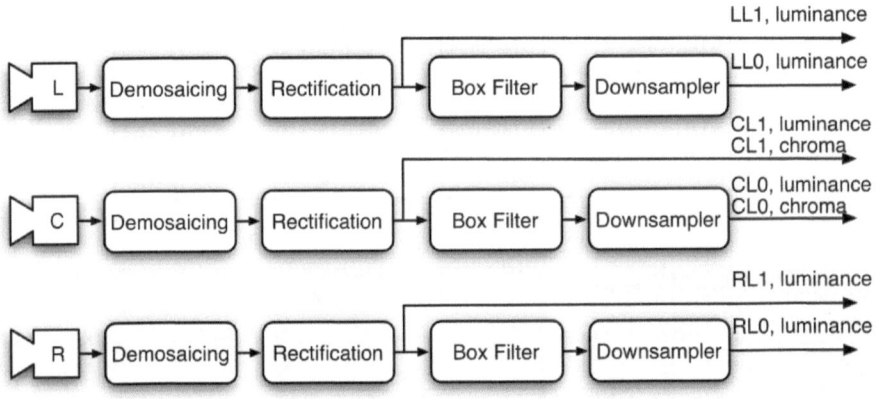

Fig. 10. Pre-processing module

Pixels generated by the camera are formatted in a Bayer pattern consisting of the four colors: Red (R), Green1 (G1), Blue (B) and Green2 (G2), representing the three color filters. The high quality linear interpolation demosaicing algorithm [13] is used to estimate the color components for each pixel.

The proposed architecture makes use of the YC_BC_R color space. The Luminance (Y) values are used to compare the two input streams. While the chrominance values (C_B, C_R) are used to construct the binary support window. Hence, the reconstructed RGB color space needs to be transformed into the YC_BC_R color space (14).

$$\begin{cases} Y = 16 + (66 \cdot R + 129 \cdot G + 25 \cdot B) \\ C_B = 128 + (-38 \cdot R - 74 \cdot G + 112 \cdot B) \\ C_R = 128 + (112 \cdot R - 94 \cdot G - 18 \cdot B) \end{cases} \quad (14)$$

Two different kinds of distortions are present in a trinocular camera setup. The first kind consists of the lens distortions; the second kind consists of the misalignment of the three cameras. Since the search space is only located on the epipolar line, both distortions should be resolved before matching can be performed. The intrinsic and extrinsic parameters of the cameras individually and the transformation matrix of the trinocular setup are determined offline using images of checkerboard patterns [14]. These parameters are hence used to construct the x and y mapping coordinates for each pixel in the image. Those parameters are called the reverse mapping coordinates.

The rectification module proposed in [15] consists of three main parts (see Fig.11). The reverse mapping coordinates are stored in a LUT. When the mapping coordinates do not change drastically from pixel to pixel, it suffices to only store the mapping coordinates of certain pixels. These pixels are chosen to be located on a regular grid. The desired grid size depends on the amount of distortion in the image. Bilinear interpolation is used to reconstruct the mapping coordinates for the complete image. The warped image is constructed by selecting the pixels from the source image whose coordinates are provided by the reverse mapping LUT. In order to perform reverse mapping, the source pixels need to be stored in an input buffer. This input buffer is especially designed to keep the memory usage low so that no external memory is needed. Third, the output pixels are resampled in order to get sub-pixel accuracy.

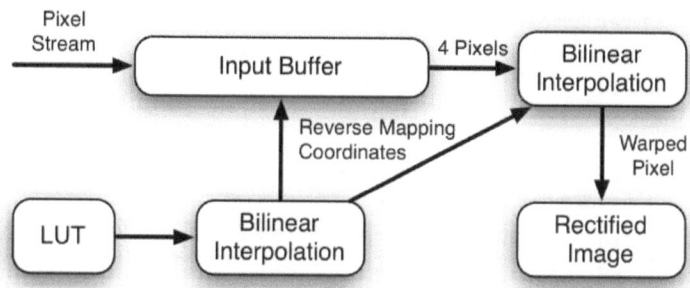

Fig. 11. Image rectification module [15]

The rectified pixel stream is passed through a 3x3 mean filter and down sampled by a factor of two. The original pixel stream is annotated with level 1 (L1) while the scaled pixel stream is annotated with level 0 (L0).

4.2   Window Comparison Module

The pixel streams originating from the right and left camera are compared with the center camera using segmentation based SAD calculation (see Fig. 12). During every clock cycle a window of the center camera is compared with four windows of the left or right camera. Since four successive pixels are stored in one memory location, one memory read accesses four pixels; hence four comparison modules are running in parallel.

On every clock cycle, the stream selection unit (SSU) determines where each data stream is written to and which windows are compared.

The frequency of the window matching module directly controls the possible disparity search width of the trinocular matching architecture and can be adapted to the available resources. The higher the frequency difference between the pixel stream and the window matching module, the more comparisons can be executed. In the example on Fig. 13 the window matching module is clocked twenty-four times higher than the pixel streams.

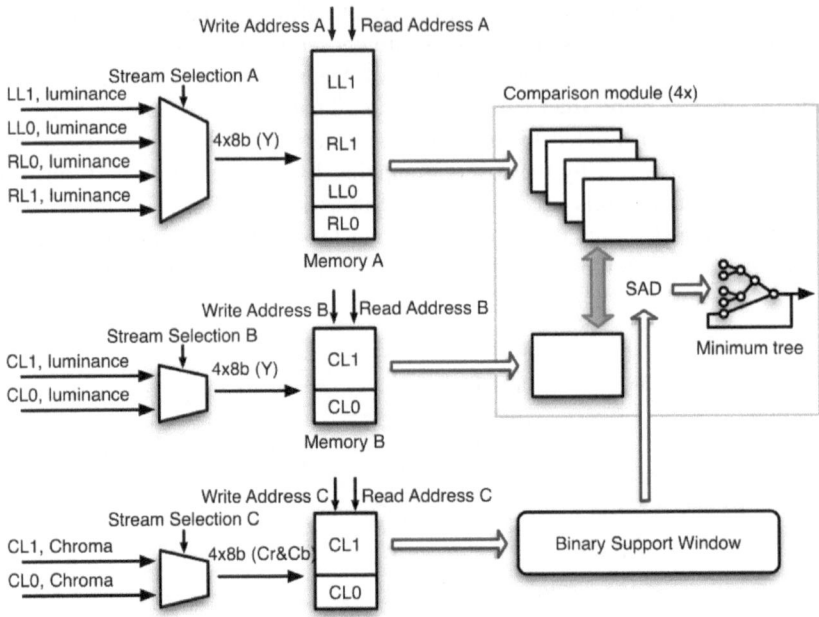

Fig. 12. Window comparison module

On each clock cycle, the comparison module compares the reference window with four consecutive windows (see Fig. 13). The lowest SAD score and its corresponding index are saved in a register, so that on the next clock cycle this lowest SAD score can be compared against the SAD scores of the next four windows. When the end of a search window is reached, the index indicates the disparity result and a new search window is initiated. In our example, in the first eight clock cycles, the center image is compared with the right image. In the next eight clock cycles the center image is compared with the left image.

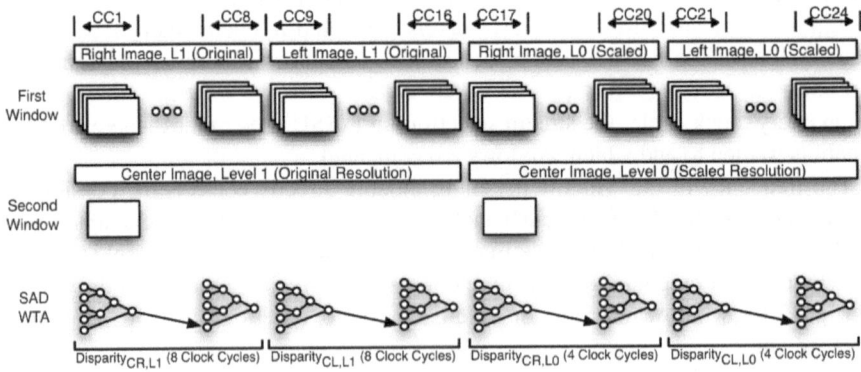

Fig. 13. Window comparison of different data streams

In the following four clock cycles the scaled center image is compared with the scaled right image and in the last four clock cycles the scaled center image is compared with the scaled left image. This leads to a combined disparity search width of thirty-two.

This architecture makes it possible to easily change the disparity search width and comparison data streams for each pixel in the DSI. By adapting the SSU it is possible to switch between a trinocular and a stereo disparity search. The trade-off is the disparity range; on each clock cycle, four comparisons can be performed. When using only two cameras, all clock cycles can be used for this camera pair. While with three cameras, only half of the clock cycles remain for each camera pair, this will lead to a reduction of the disparity search width.

4.3   Hierarchical Classification Module

The hierarchical classification module consists of the generation of the features used during the classification phase and the two classification steps. The first level classifier calculates the confidence of each stream in the selection (15, 16). For each stream, different thresholds are selected. However, the main structure of the classifier remains the same. The second level classifier selects the most promising disparity stream for the final DSI (17, 18, 19).

$$\text{streams} = \{CL0, CL1, CR0, CR1\} \tag{15}$$

$$\begin{cases} Conf_x = & if\left(SNDDBW_{21,10} < a\right)then \\ & [if\,(TEX > b)then\ 1\ else\ 0\,]\ else\ 0 \\ & x \in streams \end{cases} \tag{16}$$

$$SSDD_x = \sum_{y \in streams} Conf_y * \left|D_x \text{-} D_y\right|, x \in streams \tag{17}$$

$$Disparity\_selection = \min_{y \in streams} SSDD_y \tag{18}$$

$$Disparity = streams\,(Disparity\_selection) \tag{19}$$

The features are calculated at different moments in the streaming pipeline. For the first level of classification, three main timing zones have been specified.

- Zone 1: Features based on the luminance window e.g. TEX.
- Zone 2: Features based on the cost curve e.g. MC.
- Zone 3: Features based on the disparity window e.g. SNDD.

Features of different timing zones are expensive to synchronize. Each feature needs a buffer to temporally store its value. This is of particular importance when combining features from zone 1 and 3. Zone one features are calculated even before the calculation of the depth value, while zone three features are based on a window of disparity values around the disparity value of which the feature is calculated. This means that a delay of several image lines is to be expected.

A solution is to split the classification method into separate classification methods for each zone. For a decision tree classifier, no major changes are needed, each branch of the DT can be pre-calculated in a different zone (see Fig. 14). Since only the results need to be buffered, the width of the buffer is reduced to one.

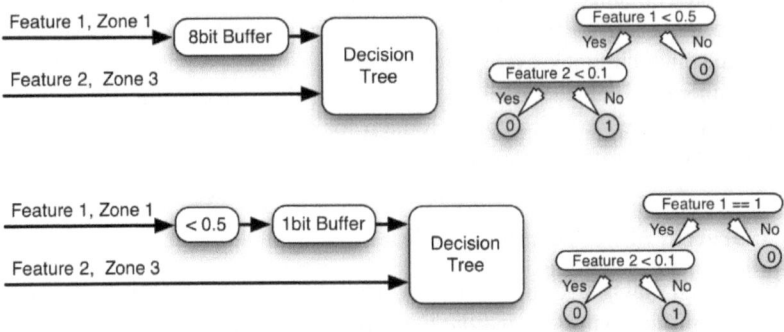

Fig. 14. Example of different timing zones for classification

5   Implementation

The architecture and methods presented in this paper have been implemented on an FPGA system (Terasic DE2-115 development board) , based on an Altera Cyclone IV (EP4CE115F29C7N) with 114,480 logic elements and 432 memory blocks. The sources of the input streams are three cameras with a resolution of 640x480 and a pixel clock of 16 MHz resulting in a refresh rate of 52 Hz. The current implementation consists of the proposed design using a 7x7 binary adaptive window SAD with a window matching clock of 96 MHz. The hardware block diagram can be found on Fig. 15.

The architecture has been constructed to reduce memory usage. Hence there is no need for external memories. The reduction of external memory usage has the additional advantage that the latency between input frame and output frame becomes minimal. This makes this system suitable to be incorporated in real-time control loops. In addition to the evaluation presented in section 2, the system has also been tested in real life environments.

The synthesis results can be found in Table 3.

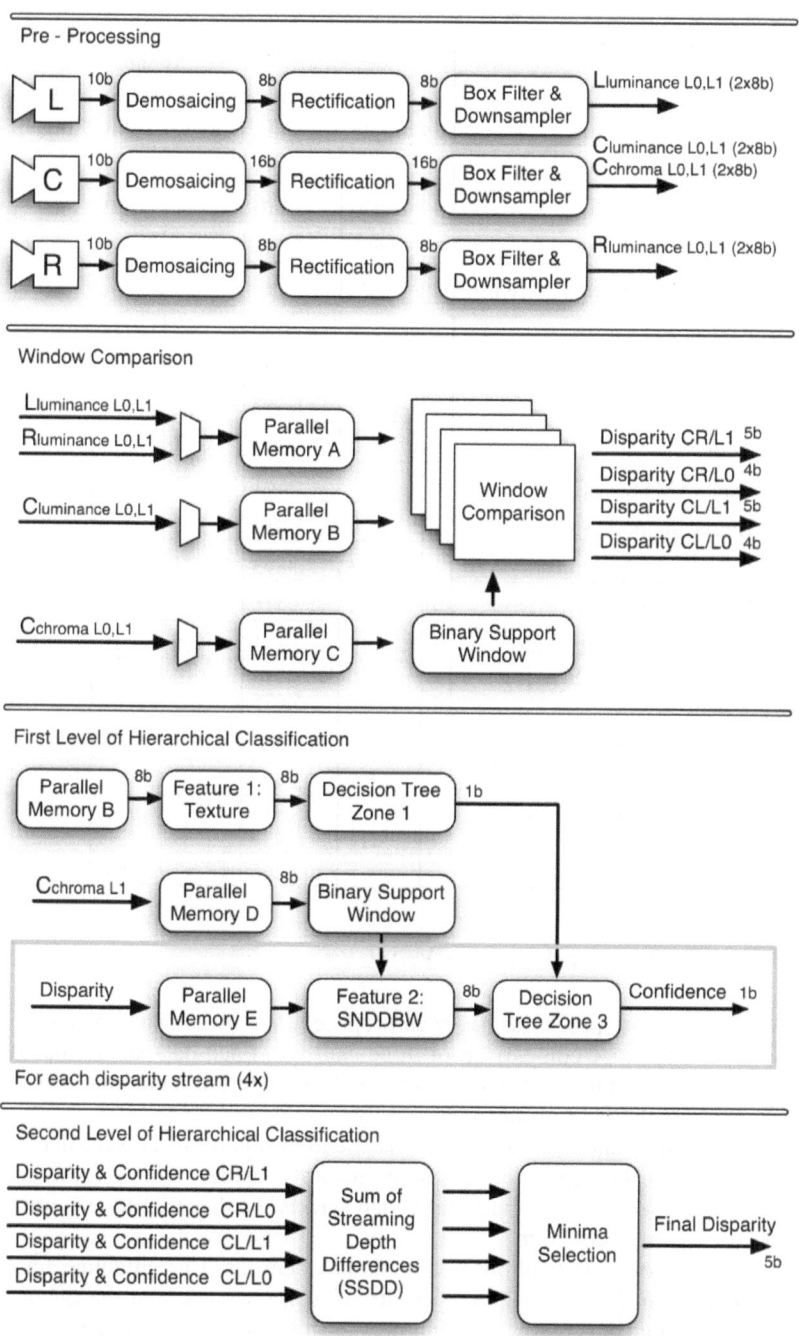

Fig. 15. Hardware block diagram

Table 3. Synthesis results overview

	Module Name	#	Logic Elements		Memory Elements			
					Nr of Blocks		Kilobits	
			Single	Total	Single	Total	Single	Total
Pre-Processing	Demosaicing	3	861	2,583	6	18	31	92
	Rectification	3	4,870	14,610	44	132	331	993
	Box Filter & Downsampling	3	688	2,064	5	15	28	84
Window Matching	Address & Stream Selection A	1	296	296				
	Parallel Memory A	1	659	659	22	22	119	119
	Address & Stream Selection B	1	340	340				
	Parallel Memory B	1	1,557	1,557	22	22	59	59
	Address & Stream Selection C	1	310	310				
	Parallel Memory C	1	661	661	22	22	59	59
	Binary Support Window	1	5,014	5,014				
	Comparison Module	1	20,893	20,893	1	1	0	0
Hierarchical Classification	Feature 1: Texture	4	854	3,416				
	Feature 2: Parallel Memory D	1	2,115	2,115	18	18	102	102
	Feature 2: Binary Suppport Window	1	5,675	5,675				
	Feature 2: Parallel Memory E	4	2,115	8,460	12	48	64	255
	Feature 2: SNDBBW (21x21)	4	8,350	33,400				
	Feature 3: SSDD & Minima Selection	1	212	212				
Total:				**102,265**		**298**		**1,764**

6   Conclusions and Future Work

A trinocular disparity processor has been proposed. We investigated nine cost curves resulting from pairwise comparison of three cameras. Each data stream has been investigated independently from one another and ultimately a hierarchic classification algorithm selects the most promising disparity value.

For each of the nine cost curves, a classification algorithm is trained in order to provide a confidence indication for their disparity values. These confidences are passed on to the second level classifier which selects the disparity to use, or indicates that no disparity has been found.

The selection of classification algorithms has been used as guideline for the implementation in an FPGA. From the results we can conclude that the quality of the disparity space image increases by using more cost curves from a trinocular camera.

Due to the adaptability of the window matching module and the hierarchic classification structure, the system can easily be expanded with more data streams to further improve the disparity space image.

References

1. Mozerov, M., Gonzalez, J., Roca, X., Villanueva, J.J.: Trinocular stereo matching with composite disparity space image. In: 16th IEEE International Conference on Image Processing, Proceedings IEEE ICIP 2009, pp. 2089–2092 (2009)
2. Ueshiba, T.: An efficient implementation technique of bidirectional matching for real-time trinocular stereo vision. In: 18th International Conference on Pattern Recognition, Proceedings IEEE ICPR 2006, pp. 1076–1079 (2006)
3. Yoon, K.J., Kweon, I.S.: Adaptive support-weight approach for correspondence search. IEEE Trans. PAMI 28(4), 650–656 (2006)
4. Motten, A., Claesen, L.: A Binary Adaptable Window SoC Architecture for a Stereo Based Depth Field Processor. In: Proceedings IEEE VLSI-SOC 2010, 18th IEEE/IFIP International Conference on VLSI and System-on-Chip, Madrid, September 27-29, pp. 25–30 (2010)
5. Hu, X., Mordohai, P.: Evaluation of stereo confidence indoors and outdoors. In: Proceedings IEEE CVPR 2010, 23rd IEEE Conference on Computer Vision and Pattern Recognition, pp. 1466–1473 (2010)
6. Motten, A., Claesen, L., Pan, Y.: Binary confidence evaluation for a stereo vision based depth field processor SoC. In: Proceedings IEEE ACPR 2011, 1st Asian Conference on Pattern Recognition, Beijing, November 28-30, pp. 456–460 (2011)
7. Jin, S., Cho, J., Pham, X.D., Lee, K.M., Park, S.-K., Jeon, J.W.: FPGA Design and Implementation of a Real-Time Stereo Vision System. IEEE Transactions on Circuits and Systems for Video Technology 20(1), 15–26 (2010)
8. Motten, A., Claesen, L.: Low-cost real-time stereo vision hardware with binary confidence metric and disparity refinement. In: Proceedings IEEE ICMT 2011, International Conference on Multimedia Technology, pp. 3559–3562 (2011)
9. Li , M., Jia, Y.: Stereo vision system on programmable chip (SVSoC) for small robot navigation. In: Proceedings IEEE/RSJ IROS 2006, International Conference on Intelligent Robots and Systems, pp. 1359–1365 (2006)
10. Scharstein, D., Szeliski, R.: A taxonomy and evaluation of dense two-frame stereo correspondence algorithms. International Journal of Computer Vision 47(1), 7–42 (2002)
11. Scharstein, D., Szeliski, R.: High-accuracy stereo depth maps using structured light. In: Proceedings IEEE CVPR 2003, IEEE Conference on Computer Vision and Pattern Recognition, pp. 195–202 (2003)
12. Hirschmüller, H., Scharstein, D.: Evaluation of cost functions for stereo matching. In: Proceedings IEEE CVPR 2007, International Conference on Computer Vision and Pattern Recognition, pp. 1–8 (2007)
13. Malvar, H., He, L., Cutler, R.: High-Quality Linear Interpolation for Demosaicing of Bayer-Patterned Color Images. In: Proceedings IEEE ICASSP 2004, IEEE International Conference on Acoustics, Speech, and Signal Processing, May 17-21, pp. 485–488 (2004)
14. Zhang, Z.: Flexible Camera Calibration by Viewing a Plane from Unknown Orientations. In: Proceedings IEEE ICCV 1999, 7th IEEE International Conference on Computer Vision, Kerkyra, September 20-25, pp. 666–673 (1999)
15. Motten, A., Claesen, L., Pan, Y.: Adaptive Memory Architecture for Real-Time Image Warping. In: Proceedings IEEE ICCD 2010, 30th IEEE International Conference on Computer Design, Montreal, September 30-October 3, pp. 466–471 (2010)

Spatially-Varying Image Warping: Evaluations and VLSI Implementations

Pierre Greisen[1,2], Michael Schaffner[1,2], Danny Luu[1], Val Mikos[1], Simon Heinzle[2], Frank K. Gürkaynak[1], and Aljoscha Smolic[1,2]

[1] ETH Zurich, Switzerland
[2] Disney Research Zurich, Switzerland

Abstract. Spatially-varying, non-linear image warping has gained growing interest due to the appearance of image domain warping applications such as aspect ratio retargeting or stereo remapping/stereo-to-multiview conversion. In contrast to the more common global image warping, e.g., zoom or rotation, the image transformation is now a spatially-varying mapping that, in principle, enables arbitrary image transformations. A practical constraint is that transformed pixels keep their relative ordering, i.e., there are no fold-overs. In this work, we analyze and compare spatially-varying image warping techniques in terms of quality and computational performance. In particular, aliasing artifacts, interpolation quality (sharpness), number of arithmetical operations, and memory bandwidth requirements are considered. Further, we provide an architecture based on Gaussian filtering and an architecture with bicubic interpolation and compare corresponding VLSI implementations.

Keywords: image-based rendering, EWA splatting, bicubic resampling, video processing, anti-aliasing, interpolation quality, complexity evaluation, caching, VLSI.

1   Introduction

With the steadily increasing frame rates and resolutions, real-time video processing and graphics processing is becoming predominant in terms of computational requirements in mobile devices. Many application-specific hardware cores for video processing are currently being integrated onto mobile system-on-chips (SoCs) (e.g., NVIDIA Tegra). One upcoming application for mobile devices is video content adaptation: while a growing amount of content is watched on an increasing number of different mobile platforms, most content is captured with one acquisition system at fixed parameters. Examples for content adaption algorithms are content-aware video resizing (video retargeting) [13], non-linear stereoscopic 3D (S3D) adaption [14], 2D to S3D conversion and S3D to multiview generation [6, 19, 5]. Other content transformation applications are camera alignment for S3D video and panoramic shots.

As a first step, any display adaptation algorithm determines an image *warping function* that is dependent on the display characteristics. The input frames

A. Burg et al. (Eds.): VLSI-SoC 2012, IFIP AICT 418, pp. 64–87, 2013.

are then transformed to the output frames according to the given warping function using a *view rendering* algorithm based on spatially-varying warping. The generation of the warping function is application-specific, and can be separated from the view rendering. For instance in video retargeting, the warping function retains the aspect ratio of *salient* (i.e., visually important) parts of the image, while the image distortion is hidden in visually less important regions. In S3D to multi-view conversion, the warping function is derived from the 3D structure of the scene (obtained from a disparity estimation step) to generate in-between views.

Various view synthesis and image rendering architectures have been proposed in prior work. However, the majority of these architectures have been optimized for one particular rendering application, such as depth-image based rendering (DIBR) [11, 3], stereo rectification [7], or non-linear lens correction [2, 17]. In contrast, we consider warping with general transformations that can be used for global per-frame transformations such as (wide-angle) lens undistortion, but also for spatially varying per-pixel transformations such as in video retargeting. Due to the spatially-varying nature of the transformation and the high resolutions of video footage in current applications, special care has to be taken in algorithm and architecture design. That is, aliasing needs to be avoided, high-quality interpolation should be guaranteed, and high computational – and memory bandwidth requirements need to be addressed.

In this paper, we address hardware efficiency and VLSI architectures of non-linear warping for view synthesis applications. It is an extended version of our previous work [8, 9]: next to the *elliptical weighted average (EWA) rendering system* presented in [8, 9] we present *non-linear image warping through bicubic interpolation and adaptive super-sampling* and assess *image quality and hardware requirements* by comparing an extended set of non-linear warping strategies. An important hardware consideration is thereby the memory bandwidth requirements and the corresponding *cache simulations and VLSI designs*. Finally, using the obtained ASIC implementation results, we provide a comparison of different warping techniques in terms of VLSI performance.

2   Non-linear Image Warping

Non-linear image warping is the process of geometrically transforming an image with a general image transformation (warping) function. In the simplest case, the image warping function can be represented as a global per-image transformation such as a rotation or translation of all the pixel values. Such transformations are usually represented by simple, per-image arithmetic operations of the input pixel locations. Stereo rectification is a practical application example: two non-aligned camera images are rectified in order to eliminate any vertical offsets between the cameras, and a 3-by-3 matrix with 8 degrees of freedom is enough to specify the full image transformation.

While our setup is able to perform global per-image transformations, its strength lies in the ability to realize locally-adaptive non-linear deformation of

Initial Image   Spatially-adaptive   Linear scaling
                       aspect ratio retargeting

Fig. 1. Example of transformations possible with our non-linear image warping setup. In addition to global per-frame transformations such as rotations or linear scaling (right image) our system also allows arbitrary non-linear transformations (middle image). Such transformations are essential for spatially-adaptive retargeting applications. Image credits: the initial image (left) is in the public domain, the middle image is generated with the framework from [13] and the right image is a linear scaled version of the initial image.

the input video, which is required in modern video applications such as content-aware video retargeting. Our warping function can be specified by a per-pixel mapping function: any pixel in the source image is assigned its own destination pixel position in the target image. Figure 1 shows an example of transformations that are possible with the system presented in this work.

2.1 Warping Basics

In the following we briefly summarize the image resampling process, for a thorough derivation we refer to literature (e.g. [18, 21, 22]). Consider an input image with pixel values w_k, where k is the linearized image coordinate and $\mathbf{u}_k$ the corresponding 2D coordinate in source (image) space. Each (non-integer) source coordinate $\mathbf{u}$ is transformed via mapping $\mathbf{m}$ to a target coordinate $\mathbf{x}$

$$\mathbf{x} = \mathbf{m}(\mathbf{u}) \qquad (1)$$

Further, let $f_i()$ be a continuous source space interpolation filter and $f_a()$ a continuous target space anti-aliasing filter. The general mapping function $\mathbf{m}$ then transforms an input image into an output image f_{out} according to

$$f_{out}(\mathbf{x}) = \sum_k w_k f_i(\mathbf{m}^{-1}(\mathbf{x}) - \mathbf{u}_k) * f_a(\mathbf{x})$$

$$= \int_{\mathbb{R}^2} \sum_k w_k f_i(\mathbf{m}^{-1}(\boldsymbol{\tau}) - \mathbf{u}_k) h(\mathbf{x} - \boldsymbol{\tau}) d\boldsymbol{\tau}. \qquad (2)$$

The interpolation and anti-aliasing filter are crucial for obtaining good image quality in the resampling process, and omitting them can lead to aliasing or holes in the output image, especially for spatially-varying transformations. The final output image is obtained by evaluating f_{out} on the desired integer grid positions.

In practice, the general mapping function $\mathbf{m}(\mathbf{u})$ is linearly approximated with a first-order Taylor expansion around an integer grid position $\mathbf{u}_k$

$$\mathbf{x} = \mathbf{m}(\mathbf{u}) \approx \mathbf{m}(\mathbf{u}_k) + \mathbf{J}_k \cdot (\mathbf{u} - \mathbf{u}_k), \qquad (3)$$

where $\mathbf{J}_k$ is the 2×2 Jacobian matrix of $\mathbf{m}$ at position $\mathbf{u}_k$. Also, the resampling equation (2) can be evaluated in two ways. The so-called backward mapping approach steps through the output pixel positions $\mathbf{x}_h$ and looks for the corresponding pixels in the input image. The forward mapping approach steps through the input grid positions $\mathbf{u}_k$ and calculates its contributions to the target pixels. In the following, we introduce practical backward and forward mapping techniques for non-linear warping.

2.2   Forward Mapping: EWA Splatting

An efficient forward mapping approach is elliptical weighted average (EWA) splatting [9]. In the EWA framework, 2D Gaussian filters are used for both interpolation and anti-aliasing filters with covariance matrices $V_{\{i,a\}} = \sigma^2_{\{i,a\}} I_2$, where I_2 is a 2-by-2 identity matrix. Two main advantages of Gaussian filters make the EWA framework very effective: first, a Gaussian filter remains Gaussian under linear transformations. Second, the convolution of two Gaussian filters results in another Gaussian.

Consider an input image with pixel values (intensities or RGB components) w_k, where k is the linearized 2D image coordinate corresponding to the 2D position vector $\mathbf{u}_k$ and an Taylor-approximated mapping $\mathbf{m}(\mathbf{u}_k) + \mathbf{J}_k(\mathbf{u} - \mathbf{u}_k)$. The complete EWA resampling process is then summarized as follows. First, the per-pixel covariance matrix is calculated from the warping grid Jacobian $\mathbf{J}_k$ and covariance matrices $\mathbf{V}_{\{a,i\}}$

$$\mathbf{C}_k = \mathbf{J}_k \mathbf{V}_i \mathbf{J}_k^T + \mathbf{V}_a. \qquad (4)$$

The technique from [9] further adapts the resulting co-variance matrix $\mathbf{C}_k$ on a per-pixel level and thereby optimizes the inherent blur-aliasing trade-off of Gaussian filters (see [9] for details). Next, for each input pixel k with position $\mathbf{u}_k$ and value w_k, we accumulate its contributions in the target image v_h on target grid positions $\mathbf{x}_h$ with linear index h

$$v_h \leftarrow \frac{w_k |\mathbf{J}_k|}{2\pi \sqrt{|\mathbf{C}_k|}} e^{-0.5(\mathbf{x_h} - \mathbf{m}(\mathbf{u}_k))^T \mathbf{C}_k^{-1}(\mathbf{x_h} - \mathbf{m}(\mathbf{u}_k))}. \qquad (5)$$

The '$\leftarrow$' symbol denotes an update operation (accumulation). Due to non-idealities, a post-normalization step is necessary: v_h/ρ_h, where ρ_h are the accumulated weights

$$\rho_h \leftarrow \frac{|\mathbf{J}_k|}{2\pi \sqrt{|\mathbf{C}_k|}} e^{-0.5(\mathbf{x_h} - \mathbf{m}(\mathbf{u}_k))^T \mathbf{C}_k^{-1}(\mathbf{x_h} - \mathbf{m}(\mathbf{u}_k))}. \qquad (6)$$

In theory, $\mathbf{x}_h$ is the complete target image grid; in practice, because of the fast decay of the Gaussian kernel, the range of $\mathbf{x}_h$ can be confined by a rectangular bounding box around the transformed center of the Gaussian $\mathbf{m}(\mathbf{u}_k)$ [9]

$$\mathbf{m}(\mathbf{u}_k) + \begin{pmatrix} \pm\sqrt{\mathbf{C}_k(1,1)} \\ \pm\sqrt{\mathbf{C}_k(2,2)} \end{pmatrix}. \tag{7}$$

2.3 Backward Mapping

Backward mapping approaches do not accumulate contributions from source pixels in the target image but, conversely, perform a direct look-up for each target pixel in the source image. In order to evaluate the analytical resampling expression without using Gaussian filters, the anti-aliasing filter is usually replaced by a practical anti-aliasing technique. The resampling expression simplifies to an interpolation in source space. There exist a variety of different filter kernels that can be used for the interpolation, such as nearest-neighbor, bilinear, bicubic, or windowed-sinc interpolation kernels. In the quality evaluation section, the performance of the interpolation kernels is evaluated and compared numerically to each other.

The general backward evaluation formula can be written as

$$\rho_h = \sum_k w_k f_i(\mathbf{m}^{-1}(\mathbf{x_h}) - \mathbf{u}_k). \tag{8}$$

For instance, in the simple case of nearest-neighbor interpolation, the expression becomes $\rho_h = w_{k'}$ with $k' = \operatorname{argmin}_k |\mathbf{m}^{-1}(\mathbf{x}_h) - \mathbf{u}_k|^2$. Expressions for other interpolation filters can be derived similarly or looked-up in literature [21].

To add anti-aliasing on top of the practically efficient backward interpolation technique, different approaches exist. One is super-sampling and decimation, where image values are looked-up on a higher-resolution output grid and then decimated again to the actual required resolution. The decimation filter thereby serves as anti-aliasing filter. Another technique is mip-mapping, which is used in the texture mapping stage in current GPUs [1]. Mip-mapping keeps multiple resolution of the same image and, during look-up, uses the resolution that corresponds to the local downscale/upscale factor.

3 Evaluations

The general resampling framework described above allows many practical realizations, in particular when selecting interpolation filters and the anti-aliasing method. In this section we compare several common methods in terms of visual quality, and, more importantly, in terms of computational complexity. Finally, we evaluate memory accesses and design a cache to reduce memory bandwidth for non-linear image domain warping applications.

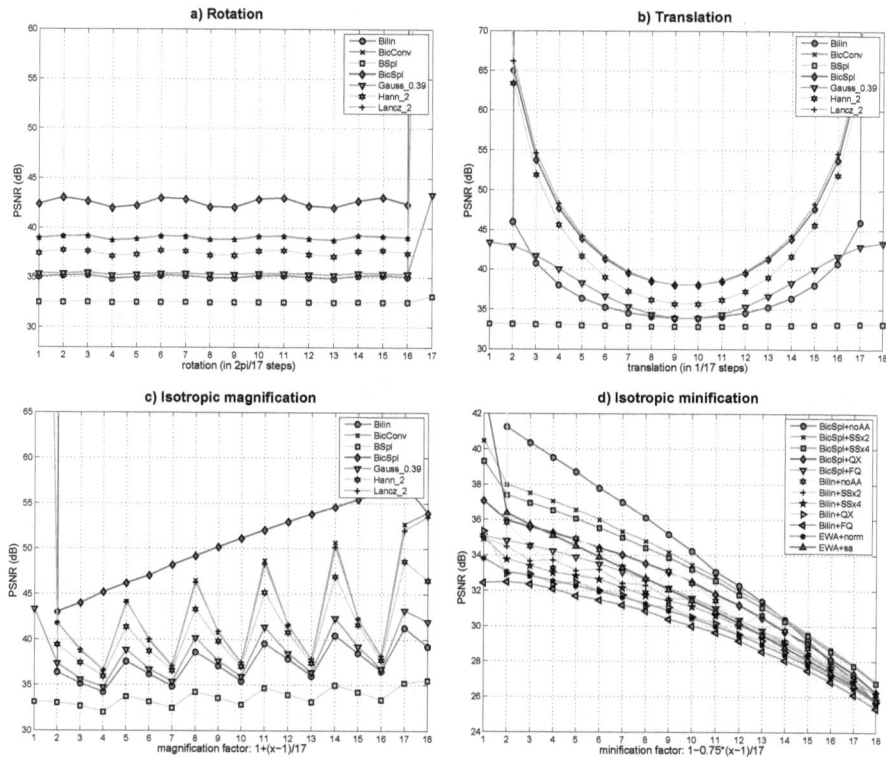

Fig. 2. Comparison of PSNR values for different kind of transformations. The evaluations have been performed by first applying a transformation and then the corresponding inverse transformation. The resulting image is compared with the original using the PSNR metric. The results plotted here are median values over a set of 16 natural 720p color images. The Gaussians are parameterized with $\sigma = 0.39$ in the above evaluation (this value was determined in [9]) and they are clipped at 2.

3.1   Quality Comparisons

For the evaluations we use several well-known interpolation and anti-aliasing methods and apply them to aspect-ratio retargeting and stereo-to-multiview conversion examples. Interpolation kernels compared in this work are bilinear and bicubic interpolation, b-spline, bicubic spline interpolation, Gaussian, and windowed-sinc filters with Hann and Lanczos windows. For details on the different methods refer to e.g. [21]. Anti-aliasing kernels in forward mapping are evaluated using the EWA framework. Backward mapping methods are evaluated with various degrees of constant super-sampling (SS) or different sampling patterns such as quincunx and flipquad [1].

Figure 2 provides evaluation results on comparing interpolation filters and anti-aliasing methods in resampling applications. The different resampling methods are used to transform a set of 16 natural 720p color images according to

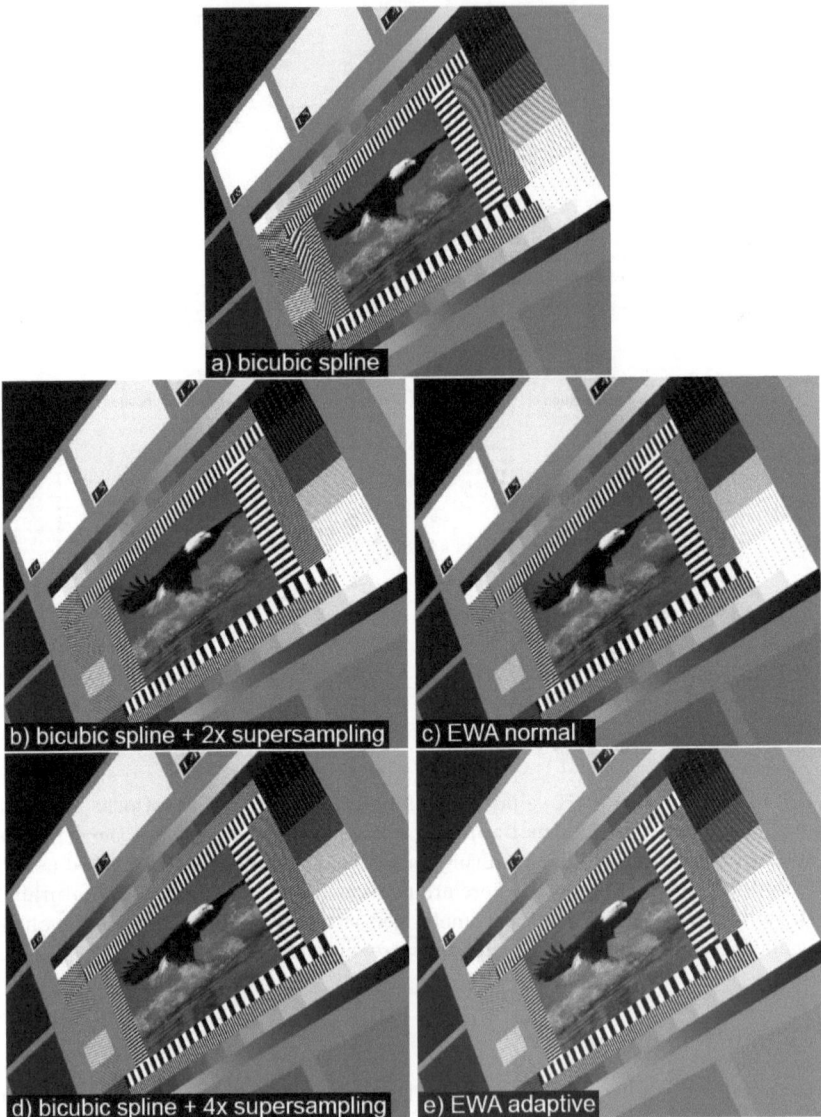

Fig. 3. Visual comparison of anti-aliasing artifacts: without anti-aliasing visual distortions are obvious (a); twofold oversampling removes some of the aliasing in this example (b). However, fourfold oversampling is necessary to remove all aliasing (d). The two EWA splatting variants (c,e) hardly show aliasing artifacts. We also observe that adaptive EWA splatting (e, see Section 2.2) provides a sharper image than conventional EWA splatting (c).

simple parameterized transformations (e.g. rotation by a certain angle). In order to be able to assess the resampling quality using the PSNR measure, the images are resampled twice: once with the forward transform and once with the corresponding inverse transform. The original images can then be used as a reference. It should be noted that the PSNR is a good measure for the amount of introduced blurring, but it does not capture aliasing artifacts very well as can be seen in Figure 2d, which shows the numerical results for isotropic minification.

From these basic quality evaluation figures, several observations can be made. The first observation is that Gaussian interpolation shows similar quality compared to the typically used bilinear interpolation, even without any non-linear warping. Bicubic methods and windowed sinc filters are in general superior to Gaussian and bilinear interpolation, and B-splines show the worst performance.

Regarding anti-aliasing, one observes two things: the *numerical* evaluations show that all methods introduce a similar amount of additional blurring, except of course, when using no anti-aliasing method. The adaptive Gaussian shows less blurring than the fixed-width Gaussian, which has been already shown in Section 2.2. On top of that, *visual* evaluations show that using no anti-aliasing filter may introduce severe aliasing artifacts, as can be seen in Figure 3.

Our observations are in line with claims and evaluations from previous works [1, 10, 18, 20]. Quality-wise, we conclude that the adaptive EWA splatting approach is superior both to simple bilinear interpolation and bilinear interpolation with super-sampling. In terms of quality, a bicubic or windowed-sinc backward mapping approach with (sufficient) supersampling are superior to EWA splatting.

3.2 Computational Complexity

Beside the quality evaluations, we compare the computational complexity of bilinear and bicubic backward mapping as well as adaptive EWA splatting in image domain warping applications (e.g., aspect ratio retargeting). We consider the following scenarios, first an identity transformation of a 1920×1080 image, second linear scaling from 1440×1080 to 1920×1080, and third linear scaling from 1920×1080 to 1440×1080.

Using these three cases, the order of magnitude of the complexity can be estimated. More importantly, these cases allow us to compare the *relative* complexity of different methods. As can be seen in Figure 4, the lowest-cost technique is bilinear interpolation (*Bil+noAA*), followed by bicubic interpolation (*Bic+noAA*) and adaptive EWA (*adEWA*). We also see that supersampling significantly increases the complexity of the backward mapping methods. Note that the bicubic polynomials and the Gaussian kernel evaluations are approximated with linear interpolation between lookup table values. Further, the inverse square-root is calculated using the fast inverse square-root approximation [16].

Warp Interpolation and Inversion. The numbers discussed above hide a practical issue: depending on the application, warp information is available in either forward or backward format and the conversion from one format to another requires additional computations. In typical image domain warping applications, warps are available in forward format. This has consequences for backward mapping approaches, where an additional inversion step becomes necessary (denoted

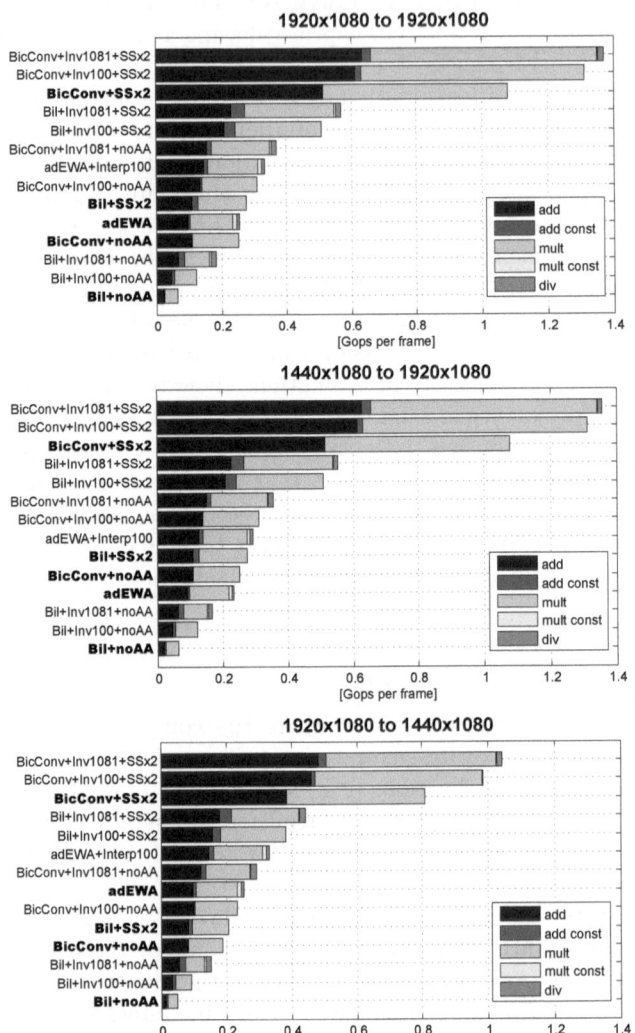

Fig. 4. Comparison of the computational complexity of different resampling techniques. Top: identity transformation; middle: 4:3 to 16:9 retargeting using linear scaling; bottom: 16:9 to 4:3 retargeting using linear scaling. The compared methods are bilinear interpolation (*Bil*), bicubic convolution (*BicConv*) and adaptive EWA splatting (*adEWA*). The postfix *SSx2* stands for twofold supersampling, whereas *noAA* stands for no anti-aliasing. The methods printed with a bold font do not include any warp preprocessing, whereas the methods with the postfix *WarpInv1081*, *WarpInv100* or *Interp100* include an additional warp inversion or warp interpolation step (the number denotes the vertical warp resolution - refer to the text for more details).

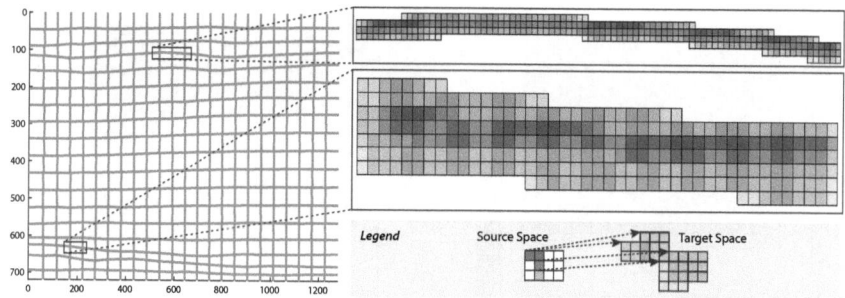

Fig. 5. This figure shows a typical frame buffer access pattern of a retargeting warp rendered with EWA splatting. The grid on the left hand side shows a typical retargeting warp of a 1280×720 image. In this application, the warp contains only limited image distortions. The excerpts on the right side show close-ups of two partially rendered regions of this warp, where the large spatial overlap among subsequent patches in horizontal and vertical directions can be observed.

as *Inv1081* in Figure 4). Further, the warp is usually available on lower resolution than the actual image such that an additional upscaling step using interpolation is necessary (e.g. bilinear interpolation). Warp inversion together with warp upsampling from 100 pixel to 1080 pixel in vertical direction is denoted as *Inv100* whereas, warp upsampling alone is denoted as *Interp100*. Note that the warp inversion also requires more memory (and associated bandwidth) for the storage of the inverted warp coordinates or the partly rasterized output image.

Conclusion. Together with the quality evaluations from the previous section, we conclude that adaptive EWA splatting forms a good tradeoff between computational complexity, interpolation and anti-aliasing quality. Bilinear interpolation lowers the computational complexity but at the price of lower image quality. Finally, bicubic approaches provide more interpolation quality at a similar cost as EWA splatting, but they are very costly when super sampling is added. In image domain warping applications, forward warp coordinates are available, which makes EWA splatting an efficient solution for non-linear warping.

3.3   Memory Bandwidth Evaluation

In real-time VLSI implementations, the warp and image are processed in scanline order. Depending on the warp format, i.e., backward or forward transformation, the image needs to be stored in either input buffer or an output framebuffer, in order to support arbitrary image transformations. The buffers usually contain large parts or the entire image and therefore have to be stored in external off-chip memories (an uncompressed full HD image amounts to about 50 Mbit). Due to the limited pin and power budget it is important to minimize the required memory bandwidth.

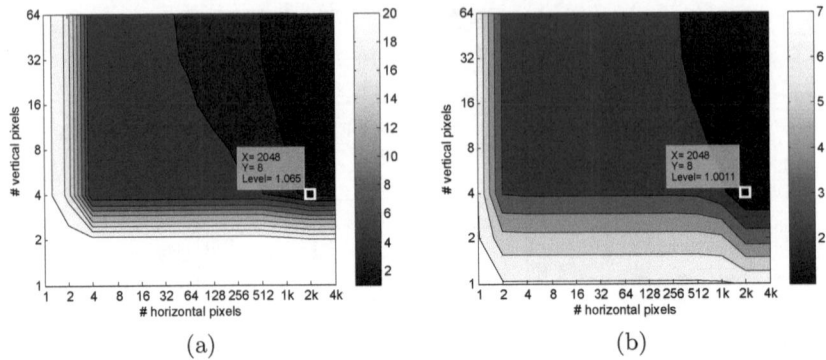

Fig. 6. Simulation of different overall cache sizes of a direct mapped cache with degenerate 1×1 blocks (i.e. each pixel has its tag and valid bit). Both plots show the averaged results from 17 nonlinear retargeting warps (aspect ratio change from 4:3 to 16:9 of 1080p images).(a) shows the normalized bandwidth of the input image buffer for bicubic backward mapping; and (b) shows the normalized bandwidth of the frame buffer for (non-adaptive) EWA splatting. The normalized bandwidth is encoded in the color. Depending on the mapping direction of the method, the normalization factor is either the size if the input (a) or output image (b).

Bandwidth Bottleneck. In the case of backward mapping, each rendered output image pixel requires a small patch to be fetched from the input image in order to perform the interpolation. Similarly, in the case of a forward mapping method, each input image pixel leads to a patch of pixel contributions in the output image, which are then accumulated in a frame buffer. In both cases the size of those patches depends on the employed filter kernel, and ranges from 2×2 (bilinear and adaptive EWA), to 3×3 (normal EWA) to 4×4 pixels (bicubic).

Thus, the amount of memory accesses is a multiple of the image resolution and can easily lead to a bandwidth bottleneck. For instance, in the case of bicubic interpolation and a 1080p color output image we require a bandwidth of $1920 \times 1080 \times 3 \times 4 \times 4 \approx 796$ MByte/frame. Fortunately, if the warp is traversed in scanline order, consecutively accessed patches spatially overlap to a certain degree as illustrated in Figure 5. The vertical and horizontal *locality* of the patches can be leveraged to reduce the overall bandwidth by employing a *two-dimensional* image cache (similar to texture caches in GPUs).

Cache Evaluations. In order to see how a two-dimensional image cache has to be parameterized in the case of image warping, we simulated different cache configurations for different resampling methods and calculated the required bandwidth for one frame. Figure 6 shows the averaged simulation results of a direct mapped cache with degenerate 1×1 blocks for 17 nonlinear retargeting warps (aspect ratio change from 4:3 to 16:9 for 1080p images). Note that the retargeting warps all show similar memory access patterns as they perform the same global linear upscaling with only local variations. For better readability, the bandwidth has been normalized with the size of the input image (a), or with the size of the

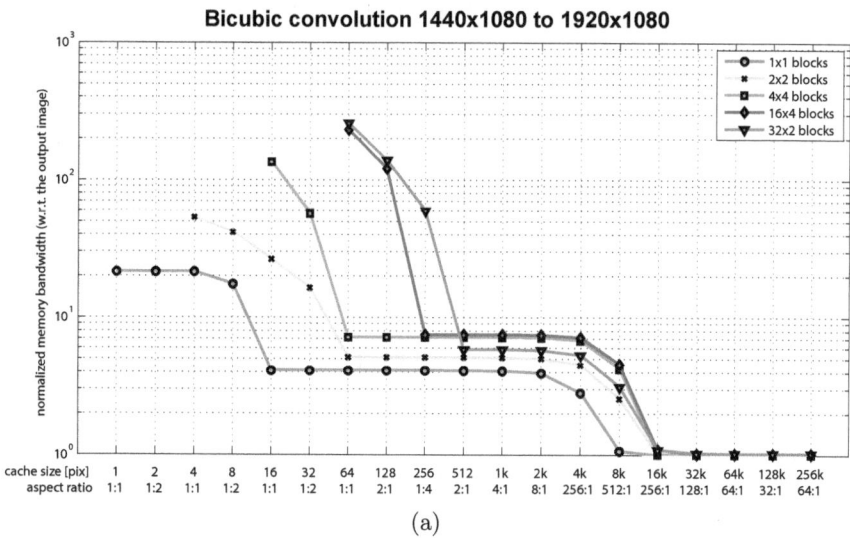

(a)

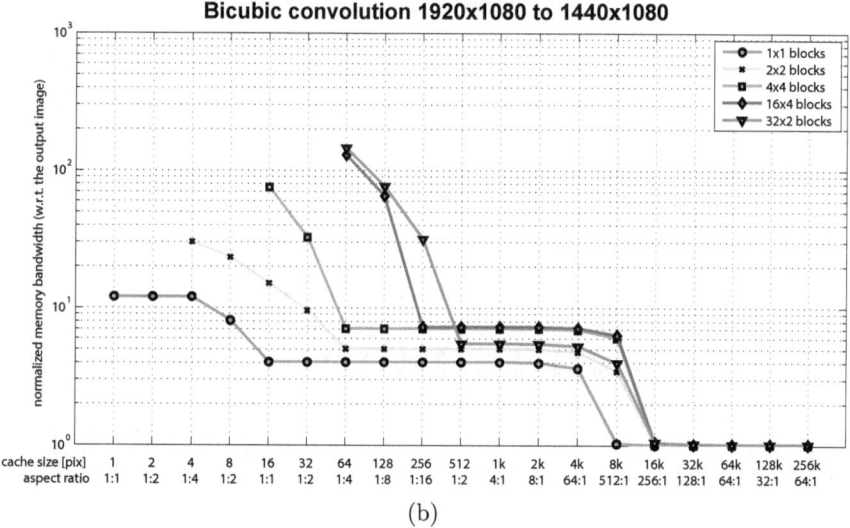

(b)

Fig. 7. Simulation of different cache configurations for the input image buffer used in bicubic backward mapping. (a) shows the results for a nonlinear aspect ratio change from 4:3 to 16:9 of a 1080p image; and (b) shows the results for a nonlinear aspect ratio change from 16:9 to 4:3 of a 1080p image. The results are mean values over 17 warps generated for natural testimages.

output image (b) – depending on the mapping direction (forward or backward). We can see that as soon as the cache is large enough to store an image patch which is larger than the size of the employed filter kernel (4×4 in (a) and 3×3 in (b)), the bandwidth is beginning to drop significantly. In the ideal case, if the cache is large enough, data will be transferred only once between the main

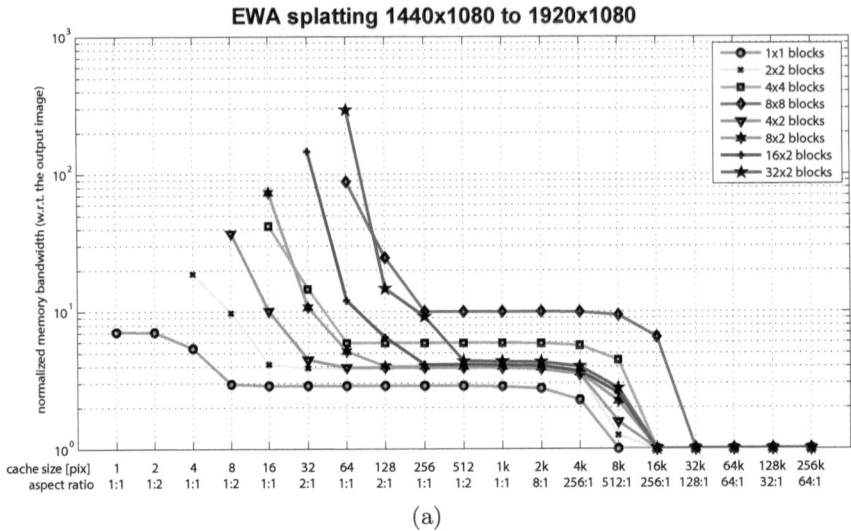

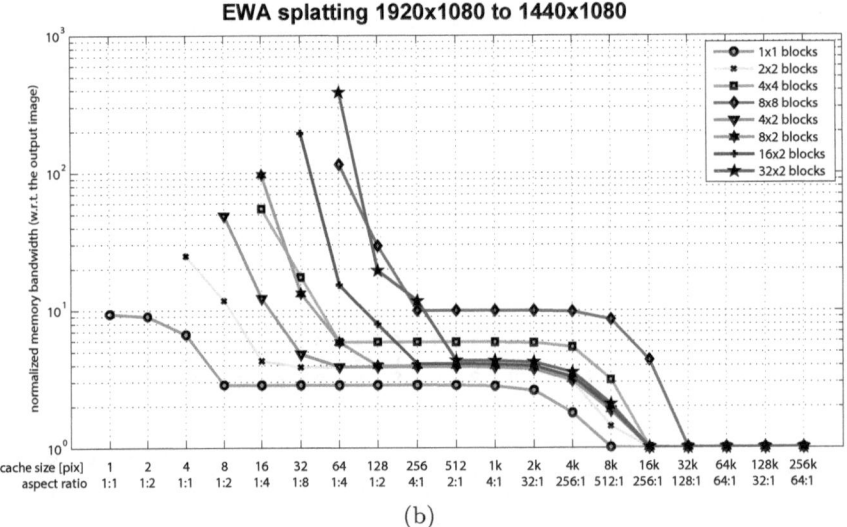

Fig. 8. Simulation of different cache configurations for the frame buffer used in EWA splatting. (a) shows the results for a nonlinear aspect ratio change from 4:3 to 16:9 of a 1080p image; and (b) shows the results for a nonlinear aspect ratio change from 16:9 to 4:3 of a 1080p image. The results are mean values over 17 warps generated for natural testimages.

memory and the cache. This corresponds to transferring only one input image to the cache in a backward mapping architecture, and only one output image in a forward mapping architecture. In this optimal case the normalized bandwidth is equal to one. As can be seen in Figure 7 and Figure 8, the optimum can be reached if the cache is large enough to hold several image rows.

A cache with a block size of 1×1 pixel is adapts well to the geometric variations in the warp function, but such a cache configuration requires a huge amount of overhead (i.e., valid-bit and address-tag entries have to be stored for each pixel as well). Increasing the block size reduces the memory required to store the tags and the valid-bits, but also comes at the cost of cache performance, as can be seen in Figure 7 and Figure 8. However, if the cache is large enough, it is possible to reach the optimum with large cache blocks.

4 Hardware Architectures

Based on the findings from the previous section, we introduce two hardware architecture for spatially-varying image warping. The first architecture uses forward mapping with adaptive EWA, the second architecture uses backward mapping with bicubic interpolation and, to reduce computations, with an *adaptive super-sampling* technique.

4.1 EWA Splatting Architecture

The top-level diagram of the EWA architecture is given in Figure 9. The core accepts streaming pixel color information, given in an 24 bit RGB format. In addition to the color information, a deformation grid describing the pixel mapping m is streamed in. In the quadrilateral deformation grid format, the deformation of each pixel is described by transformation of the pixel's bounding box. More specifically, the four corner positions of a *quad* describe the new pixel center as well as the pixel deformation. The horizontal and vertical gradients necessary for constructing J_k can be easily deduced from the quads.

We assume that the image transformation m is locally smooth, and that neighboring pixels share their adjacent quad grid corners. Therefore, in compact form, the quad grid representation only requires $(W+1) \times (H+1)$ grid points, if W and H are the input video width and height, respectively. Note that we chose this representation in order to disallow transformations that would result in image holes. Furthermore, since neighboring grid points and pixels are typically strongly correlated, we add a lossless differential compression/decompression scheme at the input interface to reduce the input bandwidth and I/O power. Note that temporal compression across frames could further reduce the input bandwidth since the warp typically varies slowly over time.

From the input quad grid, the pixel position $m(\mathbf{u}_k)$ (mean of adjacent corner positions) and the Jacobi matrix J_k (horizontal and vertical gradients computed from the corner positions) are calculated and stored in an on-chip FIFO buffer. A dispatcher unit then distributes positions, Jacobian, and pixel values to multiple arithmetic units that perform the splatting operation. The processing time of each splatting operation strongly depends on the deformation, as one input pixel can possibly be stretched to multiple output pixels. To handle the variable throughput requirements, several arithmetic splatting chains are used in parallel, and the dispatcher unit distributes the input pixels depending on the

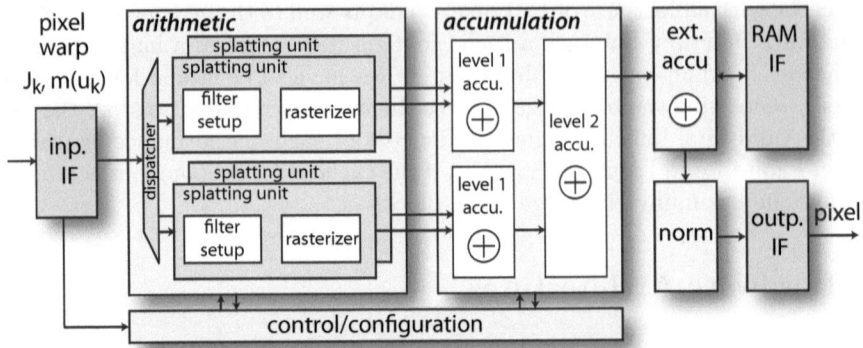

Fig. 9. Top level block diagram of the EWA rendering architecture

workload. A FIFO buffer can absorb incoming pixels when all splatting units are occupied during performance peaks. To handle prolonged peaks, the FIFO fires a back-pressure system that allows to stall the data source to avoid data loss.

Due to mathematical properties of the summation operation in (5), we can re-arrange the operation: instead of evaluating the sum for each output position, we forward-transform all individual Gaussian kernels and perform an accumulation of the Gaussian *contributions* in the temporary output image. To avoid excessive memory bandwidth requirements between the chip and the external frame buffer, we employ a two-level cache structure, in accordance with the cache configuration simulations presented in the previous section. The cache exploits the spatial coherence of image transformations, which in general map neighboring input pixels to neighboring output pixels.

When all input pixels have been processed, the temporary output image can be streamed to a normalization unit, where the accumulated pixels are then normalized by the sum of the filter weights. Note that this final normalization step is necessary due to the fact that Gaussian filters, and in particular their truncations, are non-ideal interpolation filters.

Input Interface. The system requires the pixel information and the deformation grid as the input. Since the deformation grid has to be determined by another computation block, a simple custom interface has been designed that can easily be adapted for different applications.

Compression. The input interface consists of 24 bit RGB values and 2×24 bit pixel coordinates, resulting in a bandwidth of 4.5 GBit/s for 1080p30. To reduce the input bandwidth, we propose an optional differential compression scheme. The compression exploits the fact that neighboring pixel colors and coordinates usually exhibit strong spatial correlation, and will therefore result in small incremental changes only. The purpose of the compression is to transmit the small incremental changes only. The input fixed-point words are decomposed into several sub-words, i.e., an n bit word is decomposed into n/m m bit words.

This decomposition relies on the observation, that the upper bits (MSBs) of pixels and pixel positions change very rarely compared to the lower bits (LSBs). With this, as only sub-words that change are transmitted, one MSB sub-word can be transmitted with several LSB sub-words plus control bits that indicate the number of lower sub-blocks per upper sub-block. Evaluation on actual data has shown a bandwidth reduction of 35% on average. Note that the compression is completely lossless and comes at negligible hardware overhead.

Dispatcher. The dispatcher unit is responsible for load-balancing between multiple subsequent splatting units. A simple round-robin based priority scheme is used for the scheduling.

Arithmetic Processing Elements. In a nutshell, the splatting units implement the EWA equation (5) in a fixed-point format. For each input pixel, a Gaussian kernel is calculated from the pixel color w_k and the linearized approximation J_k of the warp grid. The Gaussian kernel is then resampled to determine its contribution to all output pixels. The resampling is evaluated within a small bounding box of the Gaussian only, i.e., the Gaussian will be truncated to zero as soon as its energy falls below a very small threshold. The contributions of the individual Gaussians are then accumulated, and finally normalized.

The datapath is implemented using custom fixed-point arithmetic. The accumulated color channels are calculated with 11 bits each, and the accumulation values are calculated with 12 bits, resulting in data words of 45 bits in total for each pixel. This number has been chosen both for accuracy reasons as well as to match the word-width of the external memory.

Adaptive EWA. The splatting cores can be configured to work in 'adaptive' mode, which means that the Gaussian resampling covariance matrix is adapted per-pixel to reduce the amount of blurring. The adaptive mode has been introduced in [9] and its impact on overall area is negligible.

Throughput. Each splatting unit has a fixed throughput $\Theta = f/n_{\text{cycles}}$, determined by the clock frequency f and the number of cycles required to evaluate one input pixel n_{cycles}. The current architecture is optimized for $n_{\text{cycles}} = 20$, which is matched to the average number of output pixels times the number of cycles it takes to evaluate one output pixel (9×2 plus overhead). A throughput of 9 MPixels/s per splatting unit at a clock frequency of 170 MHz can be achieved. Therefore, 1080p30 video (63 MPixels/s) can be achieved with some margin by employing 8 parallel splatting units.

Accumulation, Caching, and Memory Interface. Each input pixel produces several output contributions that need to be weighted by a Gaussian kernel and accumulated at the output sampling locations. As described earlier, the Gaussian kernels are truncated at the bounding box boundary, and simulations have shown that bounding boxes of 4 to 9 pixels in size are sufficient to capture the majority of the non-zero contributions. Hence, the accumulation bandwidth

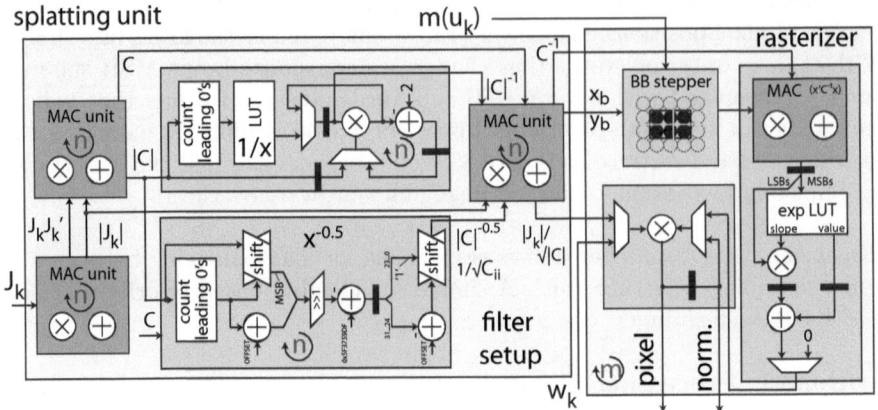

Fig. 10. Data path of the EWA splatting core

is between 2×4 and 2×9 times larger than the input bandwidth, as each accumulation is performed using a read-modify-write operation. To reduce the external bandwidth, our on-chip caching architecture exploits the horizontal and vertical overlaps of neighboring Gaussian kernels. In a first stage (denoted L1), contributions with spatial proximity are collected and accumulated into larger blocks. The L1 blocks are then efficiently accumulated to a second stage (L2 blocks). The L2 cache is able to store several lines of the image, and once a line is removed from the L2 cache it is accumulated to the external frame buffer memory. Our two-stage caching architecture reduces the resulting bandwidth considerably: the L1 cache is implemented using register arrays that support the highest bandwidth, and the L2 cache implemented using block RAMs that reduce the bandwidth to external memory further.

Throughput. Each accumulated data word has 45 bits, the required bandwidth for 1080p30 can be calculated as

$$bw_{\text{full}} = 45 \times 1920 \times 1080 \times 30 \times 2 \times (1 + n_{\text{pps}}),$$

where the factor 2 comes from the read-modify-write operation. n_{pps} denotes the number of pixels per splat, i.e., the bounding box size. Additionally, the final read out requires one more read from the memory. If we assume a conservative value of $n_{\text{pps}} = 9$, the overall bandwidth equals $bw_{\text{full}} = 56$ Gbit/s. Our cache architecture exploits the inherent spatial overlap between neighboring pixels, and shifts the bandwidth burden to the on-chip buffers, reducing the effective n_{pps}. In simulations, a cache efficiency resulting in $n_{\text{pps}} = 3$ is always achieved, and the required bandwidth is reduced to 22.4 Gbit/s.

Due to the read-modify-write operation, we choose a QDR-type memory interface to efficiently support the accumulation. QDR memories are static RAMs that have separate read and write ports, which can be accessed in parallel. Moreover, the data is transmitted in double edge mode. A 9 bit QDR RAM port therefore has 3 Gbit/s read and 3 Gbit/s write bandwidth, at a clock frequency of 170 MHz. Our architecture employs 5 instances of such 9 bit RAM interfaces, and the resulting 45 bit memory interface matches our data word size. The overall available bandwidth therefore amounts to 30 Gbit/s.

Scheduling and Control Flow. Due to the varying bounding box sizes of the input Gaussian kernels, the run-time of the individual rendering cores is non-deterministic during operation. However, on a per-frame basis the varying per-pixel run-times are averaged out and thus approximately constant, which can be used for dimensioning the number of cores and the required memory bandwidth. Short-term fluctuations of throughput are then regulated using a back-pressure system. Moreover, an efficient scheduling strategy distributes the input pixels to individual rendering units.

Output Interface. The final step of the rendering pipeline consists of reading out the image from the frame buffer and interfacing it to a standard display chip. Since display interfaces must adhere to a very strict timing, the read-out from the frame buffer is always prioritized over the read-modify-write accumulation operations. In case of collision, the accumulation can be stalled via the back-pressure mentioned before. The normalization block contains a divider producing the final 24 bit RGB values, by normalizing the accumulated RGB values with their weights.

4.2   Bicubic Warping Architecture

Figure 11 provides a top-level architecture overview of backward mapping using bicubic interpolation and two-times adaptive super-sampling. The high-level architecture is conceptually similar to the EWA architecture shown in Figure 9. The backward warping grid represented by Jacobian and backward coordinate lookup values is streamed line-by-line into the warping core (no warp inversion is performed here). However, contrary to the EWA forward architecture, the pixel intensities are not streamed together with the warp grid, but are accessed through an external buffer.

Adaptive Supersampling. The adaptive super-sampling block decides for each warp input whether super-sampling is necessary or not, by detecting if the transformation is locally minifying. Minifications are detected by checking if the determinant of the Jacobian is greater than one, or the determinant of the inverse Jacobian is larger than one, depending on which format is available at the input. In the case of supersampling, the locations of the additional sampling points are linearly approximated using the Jacobian and the lookup coordinate. Note that applying super-sampling in an adaptive way has two benefits over applying super-sampling everywhere: first, the amount of computations is significantly reduced, and second, blurring due to super-sampling with non-ideal decimation kernels is avoided where no anti-aliasing is necessary.

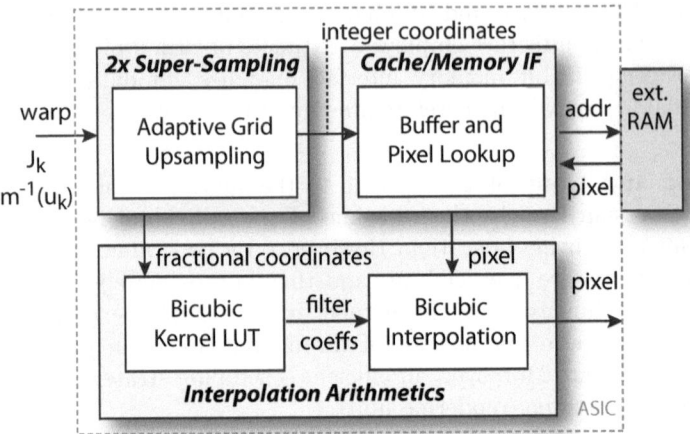

Fig. 11. Block diagram of bicubic interpolation with adaptive 2× super-sampling. The caching setup is similar to the EWA rendering architecture cache, since a similar access pattern is assumed. The key difference is that the bicubic backward mapping cache is a read-buffer whereas the EWA forward mapping cache uses (read-modify)-write accesses.

The currently employed supersampling strategy could be further improved by introducing directional super sampling, i.e., by applying super sampling only in the direction where the potential aliasing appears. This extension is able to provide a higher throughput in cases where the image transformation is demagnifying and anisotropic (e.g. an aspect ratio change from 16:9 to 4:3).

Bicubic Interpolation. The adaptive super-sampling block outputs backward coordinates to the memory interface block responsible for fetching the corresponding pixels from the external buffer. In parallel, the interpolation arithmetics block gets the fractional part of the backward coordinates to set up the bicubic filter kernel coefficients. For each output sample, an area of 4×4 pixels has to be multiplied with the bicubic kernel. Note that this kernel is separable and thus can be implemented with four vertical- and one horizontal application of the one-dimensional bicubic convolution kernel. This architecture uses fixed point arithmetic throughout. The coefficients of the bicubic kernel can therefore be directly obtained from two lookup tables (LUTs), where we the indices consist of two integer bits and the fractional bits of the x and y coordinate differences, respectively. If the fractional precision is not too large, no additional refinement of the indexed values (e.g. using linear interpolation) has to be performed as the LUTs already cover the whole index range. The throughput of the bicubic interpolator is one pixel per 4 cycles. In order to produce a super-sampled output pixel, four such samples have to be calculated and averaged and thus 16 cycles are required in that case. Thus, the effective throughput of the interpolator depends on the fraction of supersampled pixels per frame.

Caching and Memory Interface. Analogue to the forward mapping architecture, the amount of accessed input image pixels is very large as the bicubic interpolator always accesses a 4×4 neighbourhood in order to calculate an output pixel. Thus even in the case of no supersampling, we would have to load 16 times more pixels than there are in one frame, which translates into a very large external bandwidth. An on-chip read cache which is able to hold several lines of the image is therefore employed to reduce the bandwidth. As the bicubic interpolator requires 4 cycles to produce one output sample, the cache memory is divided into four column interleaved RAM macros such that four parallel accesses are possible.

5 Implementation Results

Several ASIC implementations of non-linear warping architectures have been realized which allow for a comparison and conclusion on hardware performance of the different techniques. In the following, we discuss some of the implementation results of the different ASICs.

5.1 EWA Splatting: ESPER

The EWA architecture described previously was implemented in VHDL and was fabricated in 180 nm (1P6M) CMOS technology (Figure 12(a)). The design supports image resolutions up to 2048×2048 and works on gray-level 8-bit pixels. It employs four splatting units to support 720p25 in splatting performance. Due to die size limitations the cache is reduced to eight lines of gray-valued 576p (i.e. 8x1024 pixels). The ASIC has been successfully tested at 123 MHz where a power consumption of 300 mW has been measured. Core voltage is 1.8 V and I/O pad voltage is 3.3 V. Core area is 6 mm^2 which corresponds to 660 kGE. There are 64 data I/O pins and 56 power/ground pins. This chip is a prototype of the EWA core architecture and does not possess a real-world memory interface. The normalization block is also not included. The accumulation precision is set very conservatively to 16 bit per entry.

Detailed Throughput Figures. For the following throughput figures, a nominal clock frequency of 133 MHz is assumed. One splatting unit has a throughput of 6.65 MPixels/s. A 720p25 video stream requires a throughput of 23 MPixels/s and thus four splatting units. The necessary external memory bandwidth without caching is $2 \times 9 \times 23 = 414$ MPixel/s which amounts to $414 \times 4 \approx 1.66$ GByte/s for 4 bytes per pixel entry (accumulated value plus normalization weight). The factor 2 comes from the read-modify-write operation of the accumulation. The cache has an efficiency of about 83% which reduces the external bandwidth to 282 MByte/s, i.e. the normalized bandwidth is around 1.5 (normalized to the bandwidth it takes to read, modify and write one output image). The optimum normalized bandwidth cannot be reached, as the cache is only 1024 pixels wide. In order to reach the optimal cache efficiency for 720p25 video, the cache should be extended to 8 lines with 1280 pixels per line (see Section 3.3). Note that in addition to the above bandwidth, a read/clear operation to the memory is further necessary to account for the final read-out and clearing.

5.2   EWA Splatting: VESPER

The VESPER chip is an extended version of ESPER, and it is designed to render full HD color images at 30 frames per second. In contrast to ESPER this chip has been fabricated in 130 nm CMOS and it is equipped with fully-functional display- and memory interfaces (Figure 12(c)).

The DVI interface requires a fixed input bandwidth and clock frequency, which usually is dependent on the display resolution and frame rate. To decouple the arithmetics and accumulation from the DVI interface, we separate the design into two clock domains. While the rendering core should run as fast as possible, the DVI core is running at the specific DVI pixel clock. The asynchronous data interface is implemented using an asynchronous FIFO, see [4]. The chip also provides clock signals for the external RAM components and the DVI transmitter. In order to provide a flexible timing at the corresponding interfaces, those output clocks can be phase shifted relative to the internal clock signals.

The I/O bandwidth, the throughput of the splatting units and the caching have been dimensioned for very pessimistic and demanding scenarios, such that 1080p30 performance is achieved in a practical system that supports a wide variety of warping applications. In turn, the actual performance for typical applications will be higher, and therefore also higher frame rates are possible. Under typical conditions, our architecture reaches 1080p48. Furthermore, smaller resolutions are always possible and would increase the frame rate further (e.g. 720p60).

The chip area is largely dominated by the number of input and output pins, as well as the required power distribution for the high speed I/O interfaces. VESPER supports 175 data I/O pins, of which 115 pins are used for the external QDR-II interface. Due to the prototype nature of the chip, a more conventional "around the core" I/O has been employed, instead of a more area efficient flip-chip I/O. For a commercial implementation, the area could be significantly reduced since the overall logic area (including on-chip SRAM) is $9 \, \mathrm{mm}^2$ which is much smaller than the $5 \times 5 \, \mathrm{mm}^2$ the chip currently occupies.

5.3   Bicubic Interpolation: EVA

A bicubic backward mapping ASIC named EVA has been fabricated in 180 nm CMOS technology (Figure 12(b)), and it has been designed to roughly match the specifications of the ESPER ASIC such that the two implementations are comparable. It is able to support enough throughput for 576p25 when 10% of the calculated pixels are super-sampled. Typical case post-layout simulations have shown a maximum operating frequency of 135 MHz and a power consumption of 60 mW. The core area is around $1 \, \mathrm{mm}^2$ which correspond to 110 kGE. The chip contains a similar cache configuration as ESPER (8×1024 pixel) with

Table 1. Comparison of different warping CMOS implementations. The rather large differences in size originate from the type of transformation that are supported: arbitrary (arb.) or simple linear scaling (lin. scal.). In addition, EWA and bicubic feature full or partial (super sampling (SS)) anti-aliasing support. Note that for the EWA architecture, the interpolation (interp.), anti-aliasing (AA), and transformation (transf.) blocks cannot be separated for the area numbers. A '–' means that the architecture does not contain such a block, N/A means that the values are not available. The external bandwidth figures are typical case values, normalized to reading/writing an image once. The maximum throughput of our implementations are evaluated valid for non-linear resizing (change of aspect ratio from 4:3 to 16:9). Results marked with (*) are obtained through postlayout simulations.

	EWA ESPER	Bicubic EVA	Ext. Bil. [15]	Ext. Bil. [12]
Anti-Aliasing	yes	2×SS	no	no
Transformation	arb.	arb.	lin. scal.	lin. scal.
Mapping direction	fwd.	bwd.	bwd.	bwd.
Image resolution	576p	576p	1080p	WQSXGA
Color channels	1	1	1	1
Technology [nm]	180	180	130	130
Max. clock freq.[MHz]	123	135*	267	278
Max. throughput [fps]	40	47 (0%SS)* 28 (10%SS)* 12.5 (100%SS)*	N/A	30
Power [mW]	300	62*	18.1	11.7
Transformation [kGE]	≈ 115	6	N/A	N/A
Interpolation [kGE]	≈ 115	10	N/A	N/A
Total w.o. memory [kGE]	230	16	26	13
Buffer size [Bit]	$1024 \times 8 \times 2 \times 16$	$1024 \times 8 \times 8$	–	4 lines
Buffer [kGE]	410	90	–	N/A
Memory type	DP SRAM	SP SRAM	–	SRAM
Normalized ext. BW	1	1	4	N/A

approximately half of the access bandwidth: backward mapping only requires a read operation whereas the accumulation process in forward mapping requires a read-modify-write operation. Thus, a single port memory is sufficient which reduces SRAM size by half. Further, the entries are only 8 bit wide, and no accumulation weights have to be stored. Thus, in total, the SRAM memory macro is about four times smaller in size. Note however, that this factor of four reduces to a factor of around 1.83 if RGB pixels are stored (EWA shares the weight entry among the three color subpixels), and if the accumulation precision is changed to a more realistic value of 11 bit.

5.4 Comparison

In Table 1, we list several warping VLSI architectures to compare computational resources. As can be seen, the EWA warping core is significantly larger than comparable backward architectures such as EVA and [15, 12]. In particular, the cache

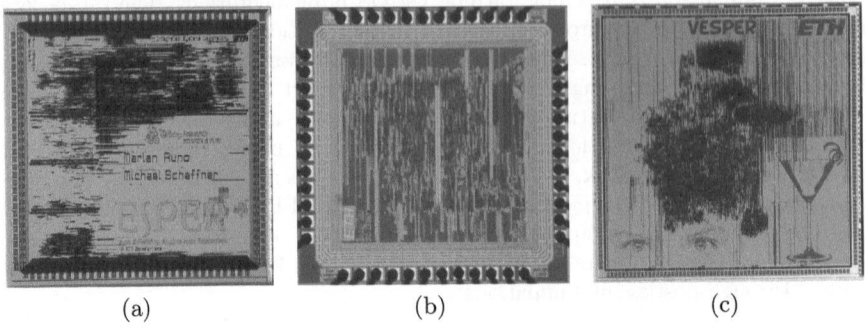

Fig. 12. Photo and CAD rendering of our VLSI implementations: (a) ESPER (180 nm), (b) EVA (180 nm) and (c) VESPER (130 nm). The pictures are not to scale.

of the EWA implementations is larger due to the higher fixed-point precision, the additional weight entry, and the read-modify-write type operation. Another reason for the higher gate count of the EWA implementations is that the EWA arithmetics are fully evaluated, whereas a significant portion of computations in the bicubic implementation are optimized by using LUT-based approximations. Note that the EWA architecture could also be optimized by replacing some of the arithmetic units with look-up tables. Thus, in applications where warping information is available in backward format and aliasing is of limited concern, one should clearly prefer a backward mapping architecture, e.g. bicubic or bilinear interpolation.

6   Conclusions

Arbitrary transformations of high-definition videos can be efficiently rendered using a non-linear warping VLSI architecture. The proposed VLSI cores can be used in an end-user device and enable image warping for current and upcoming content-adaptive applications. Due to the separation of the rendering core into several sub-units, the computational capabilities are easily scalable to higher resolutions and frame-rates, such as the upcoming quad HD standards (2160p) or the high-frame rate (HFR) standards.

References

1. Akenine-Moller, T., Haines, E., Hoffman, N.: Real-time rendering. AK Peters (2008)
2. Asari, K.V.: Design of an efficient vlsi architecture for non-linear spatial warping of wide-angle camera images. Journal of Systems Architecture 50(12), 743–755 (2004), http://www.sciencedirect.com/science/article/pii/S1383762104000682
3. Chang, F.J., Tseng, Y.C., Chang, T.S.: A 94fps view synthesis engine for HD1080p video. In: 2011 IEEE Visual Communications and Image Processing (VCIP), pp. 1–4 (November 2011)

4. Cummings, C.: Simulation and synthesis techniques for asynchronous fifo design. In: SNUG 2002 (Synopsys Users Group Conference, San Jose, CA, 2002 User Papers (2002)
5. Do, M., Nguyen, Q., Nguyen, H., Kubacki, D., Patel, S.: Immersive visual communication. IEEE Signal Processing Magazine 28(1), 58–66 (2011)
6. Farre, M., Wang, O., Lang, M., Stefanoski, N., Hornung, A., Smolic, A.: Automatic content creation for multiview autostereoscopic displays using image domain warping. In: 2011 IEEE International Conference on Multimedia and Expo (ICME), pp. 1–6. IEEE (2011)
7. Greisen, P., Heinzle, S., Gross, M., Burg, A.: An FPGA-based processing pipeline for high-definition stereo video. EURASIP Journal on Image and Video Processing 2011(1), 18 (2011)
8. Greisen, P., Emler, R., Schaffner, M., Heinzle, S., Gurkaynak, F.: A general-transformation EWA view rendering engine for 1080p video in 130 nm CMOS. In: 2012 IEEE/IFIP 20th International Conference on VLSI and System-on-Chip (VLSI-SoC), pp. 105–110 (October 2012)
9. Greisen, P., Schaffner, M., Heinzle, S., Runo, M., Smolic, A., Burg, A., Kaeslin, H., Gross, M.: Analysis and vlsi implementation of ewa rendering for real-time hd video applications. Transactions on Circuits and Systems for Video Technology (2012) (accepted)
10. Heckbert, P.: Fundamentals of Texture Mapping and Image Warping. Masters thesis, Univ. of California, Berkeley, Dept. of Electrical Eng. and Computer Science (1989)
11. Horng, Y.R., Tseng, Y.C., Chang, T.S.: VLSI architecture for real-time HD 1080p view synthesis engine. IEEE Transactions on Circuits and Systems for Video Technology 21(9), 1329–1340 (2011)
12. Huang, C.C., Chen, P.Y., Ma, C.H.: A novel interpolation chip for real-time multimedia applications. IEEE Transactions on Circuits and Systems for Video Technology 22(10), 1512–1525 (2012)
13. Krähenbühl, P., Lang, M., Hornung, A., Gross, M.: A system for retargeting of streaming video. ACM Transactions on Graphics (TOG) 28(5), 1–10 (2009)
14. Lang, M., Hornung, A., Wang, O., Poulakos, S., Smolic, A., Gross, M.: Nonlinear disparity mapping for stereoscopic 3D. ACM Trans. on Graphics (Proc. SIGGRAPH) 29(3) (2010)
15. Lin, C., Sheu, M., Chiang, H., Wu, Z., Tu, J., Chen, C.: A low-cost VLSI design of extended linear interpolation for real time digital image processing. In: International Conference on Embedded Software and Systems, ICESS 2008, pp. 196–202. IEEE (2008)
16. Lomont, C.: Fast inverse square root. Tech. rep., Purdue University (2003), http://www.lomont.org/Math/Papers/2003/InvSqrt.pdf
17. Oh, S., Kim, G.: Fpga-based fast image warping with data-parallelization schemes. IEEE Transactions on Consumer Electronics 54(4), 2053–2059 (2008)
18. Szeliski, R., Winder, S., Uyttendaele, M.: High-quality multi-pass image resampling. Tech. rep., Microsoft Research (2010)
19. Tanimoto, M., Tehrani, M., Fujii, T., Yendo, T.: Free-viewpoint tv. IEEE Signal Processing Magazine 28(1), 67–76 (2011)
20. Triggs, B.: Empirical filter estimation for subpixel interpolation and matching. In: International Conference on Computer Vision (ICCV), vol. 2, pp. 550–557 (2001)
21. Wolberg, G.: Digital image warping, vol. 3. IEEE Computer Society Press (1990)
22. Zwicker, M., Pfister, H., Baar, J.V., Gross, M.: EWA splatting. IEEE Transactions on Visualization and Computer Graphics 8(3), 223–238 (2002)

An Ultra-Low-Power Application-Specific Processor with Sub-V_T Memories for Compressed Sensing

Jeremy Constantin[1], Ahmed Dogan[1], Oskar Andersson[2],
Pascal Meinerzhagen[1], Joachim Rodrigues[2],
David Atienza[1], and Andreas Burg[1,*]

[1] École polytechnique fédérale de Lausanne, VD-1015 Lausanne, Switzerland,
Institute of Electrical Engineering
jeremy.constantin@epfl.ch
[2] Lund University, 22100 Lund, Sweden,
Department of Electrical and Information Technology

Abstract. Compressed sensing (CS) is a universal low-complexity data compression technique for signals that have a sparse representation in some domain. While CS data compression can be done both in the analog- and digital domain, digital implementations are often used on low-power sensor nodes, where an ultra-low-power (ULP) processor carries out the algorithm on Nyquist-rate sampled data. In such systems an energy-efficient implementation of the CS compression kernel is a vital ingredient to maximize battery lifetime. In this paper, we propose an application-specific instruction-set processor (ASIP) processor that has been optimized for CS data compression and for operation in the sub-threshold (sub-V_T) regime. The design is equipped with specific sub-V_T capable standard-cell based memories, to enable low-voltage operation with low leakage. Our results show that the proposed ASIP accomplishes $62\times$ speed-up and $11.6\times$ power savings with respect to a straightforward CS implementation running on the baseline low-power processor without instruction set extensions.

Keywords: Ultra-Low-Power Processor, Application-Specific Instruction Set Processor, Instruction Set Extensions, Sub-V_T Operation, Sub-V_T Embedded Memories, Compressed Sensing.

1 Introduction

Digital signal processing traditionally relies on the Nyquist sampling theorem which states that a faithful reconstruction of a signal, limited to a bandwidth B in the frequency spectrum, can be ensured with a sampling rate of $f_s \geq 2 * B$. Unfortunately, when sampled data needs to be stored or needs to be transmitted over a wireless link, the storage or transmission costs of the raw samples can often

---

[*] This chapter extends the work published earlier in [1].

A. Burg et al. (Eds.): VLSI-SoC 2012, IFIP AICT 418, pp. 88–106, 2013.
© IFIP International Federation for Information Processing 2013

limit the energy-autonomous lifetime of the system. In this case, it is advisable to first compress the data. However, in this case the power consumption of the compression process must also be kept very low to ensure an overall energy-efficiency advantage.

Compressed sensing (CS) [2] is a universal, low-complexity data compression technique to compress sparse signals. CS has been widely used in environmental monitoring systems and in wireless body sensor networks (WBSNs) [3], where portable and autonomous devices are expected to operate for long periods of time with limited energy resources. Hence, an ultra-low-power (ULP) CS implementation is crucial for these systems.

On the architectural level, supply voltage scaling, potentially all the way to the subthreshold (sub-V_T) regime, can reduce both dynamic and leakage power consumption. Therefore, many sensing platforms exploit sub-V_T computing. The state-of-the-art processors for sensing platforms have been reported to consume as little as a few pJ/cycle while operating in the sub-V_T regime [4–6]. Sub-V_T computing can also be used to perform CS data compression (in the digital domain). However, most established CS implementations either require a large memory footprint or still require considerable computational effort (despite the inherent complexity advantage of CS). Leakage power consumption becomes a very important challenge in the sub-V_T regime with reduced active power. A considerable amount of leakage in sensing platforms is due to the integrated memories [7]. Moreover, many sensing platforms cannot be power gated completely, to retain their memory content [5], and hence leakage power is always dissipated. Therefore, implementations with large memory requirements are not desirable in the sub-V_T regime. On the other hand, high computational effort requirements can limit the degree of voltage scaling because of performance degradation issues in the sub-V_T regime [8–10]. These issues ultimately limit currently the benefits of CS based data compression in ULP sensor nodes.

Application-specific instruction-set processors (ASIPs) can compensate for the performance degradation issue, since they are optimized for a specific application domain, providing increased efficiency and performance for the core algorithms of the domain's target applications. For instance, an ASIP optimized for stereo image processing can achieve up to 130× speed-up compared to a conventional processor [11]. These performance optimizations also lead to energy saving as in [12], where a processing core with few accelerators dedicated to biomedical applications, can achieve up to 11.5× energy saving compared the processing core-only implementation. Despite of their efficiency in some specialized application domains, to the authors knowledge no ASIP core has been reported for ultra-low-power CS compression.

Contributions. We propose to synergistically exploit sub-V_T computing in conjunction with an ASIP core for CS compression to provide an ultra-low-power solution for compression of sparse signals for sensing applications. To this end, we extend the instruction set of a low-power processor to exploit the specific operations of the CS compression algorithm. Our ASIP core does not require high clock frequencies, and therefore enables more aggressive sub-V_T voltage scaling for a given throughput requirement. The very low memory requirements additionally allow for a major reduction in leakage power. For a typical case study of electrocardiogram (ECG) signal compression in WBSNs, the processor consumes only 30.6 nW for an ECG sampling rate of 125 Hz. Moreover, we show that the proposed processing platform achieves 62× speed-up and 11.6× power saving with respect to the established computation-based CS implementation running on the baseline low-power processor.

2   Compressed Sensing

Signal compression based on compressed sensing (CS) [2] is performed by computing the matrix-vector multiplication:

$$y = \Phi x \tag{1}$$

where the random sensing matrix $\Phi \in \mathbb{R}^{k \times n}$ with $k < n$ maps an input data vector $x \in \mathbb{R}^n$ holding n samples to a compressed data vector $y \in \mathbb{R}^k$ with k entries, for a compression ratio of $\frac{k}{n}$.

There are multiple approaches of how to choose a random sensing matrix Φ with k rows and n columns. Sensing matrices with near optimal properties can for example be constructed by choosing the entries of Φ by random iid sampling from a uniform distribution [2].

2.1   Reduced Complexity Compression Algorithm

The structure and values of the entries of Φ determine the computational complexity of the matrix-vector multiplication. Mamaghanian et al. [3] show (for WBSNs) that in fact choosing Φ as a sparse matrix that contains only a few non-zero entries per column at random positions is a valid approach which significantly reduces complexity and still provides good integrity of the compressed sparse signals. The non-zero elements can furthermore be chosen as 1, and the number of ones per column (namely I) can be fixed. These constraints on Φ lead to a very efficient algorithm (Algorithm 1) for performing CS data compression. As a result, the computational complexity of the CS algorithm is reduced from $n \times k$ multiplications and $(n-1) \times k$ additions for a dense sensing matrix of random values, to only $I \times n$ additions. The sensing matrix can therefore be represented in a compact form by a sequence of $I \times n$ random indices $\in \{1, 2, ..., k\}$ describing for each column in Φ the rows with non-zero entries[1].

On a resource constrained system, the key challenge of Algorithm 1 is the generation of the random indices. The optimized reference implementation [3] uses a sensing matrix realized as a fixed sequence of indices stored in memory (for a specific value of k). Since large memory footprints are undesirable, especially in the context of ULP sensor nodes and sub-V_T operation, we discuss the generation of the required random indices at runtime.

---

[1] Note that strictly speaking such a representation requires unique row-indices per column. However, this requirement can often be relaxed without a significant impact.

Algorithm 1. Pseudocode of Compressed Sensing Algorithm
___
1: **for** $i := 1 \rightarrow n$ **do**
2:     $sample := getSample()$
3:     **for** $j := 1 \rightarrow I$ **do**
4:         $index := getRandomIndex(1..k)$
5:         $buffer[index] := buffer[index] + sample$
6:     **end for**
7: **end for**
___

2.2   Pseudo Random Number Generation

A pseudo random number generator (RNG) can be used for the generation of the random indices. A common implementation of such an RNG is a linear feedback shift register (LFSR). The random sequence generated by an LSFR is defined by the sequence of its internal states. The initial state of an LFSR is referred to as its seed. For each state transition (LFSR step) the current internal state bits are combined with the binary coefficients of a polynomial, which defines the pseudo random sequence of the LFSR. The bits selected by the polynomial are summed to produce one new bit (parity bit). The next state of an LFSR is calculated by shifting out the least significant bit of the state and shifting the generated bit in as the new most significant bit.

Maximum-length LFSRs provide a cycle length of the generated random number sequence that is equal to the number of maximum possible states (excluding zero). Note that although maximum-length LFSRs can provide good sequences of random numbers, the correlation between two subsequent LFSR states, i.e., subsequent indices, i_1 and i_2 is high, since $i_2 = i_1/2$ or $i_2 = i_1/2 + k/2$. When the state is used directly, this correlation of the generated indices has a negative effect on the reconstruction quality of the compressed samples. Hence, we propose to use an LFSR that advances multiple steps per generated index. The number of steps is equal to the number of used index bits. For example, for $k = 256$ the LFSR has to advance 8 states to generate the next index, which yields only a small correlation to its predecessor. The quality of our generated random indices for CS is assessed in the case study presented in Section 4.4. The drawback of this approach is the increased computational effort for the RNG, which can be compensated for by custom hardware support.

The proposed generation of the sensing matrix Φ can hence be described with four main parameters: the LFSR polynomial, the LFSR seed, the number of index bits (depending on k), and the number of non-zero elements per column (I). These four configuration parameters enable the generation of a large set of different sensing matrices. At the same time, the compact representation of the RNG configuration keeps the memory overhead small compared to the case where all indices for multiple matrices would need to be stored.

Hence, by choosing from a preconstructed pool of feasible values for the RNG configuration, it is hence possible to achieve good sensing performance for a variety of different signal conditions, potentially even by dynamically changing the RNG configuration at runtime. This capability supports one of the strength of the compressed sensing method which lies in the fact that even a randomly chosen sensing matrix Φ ensures a good mapping for the sample data of a signal source which has sparsity in a specific (potentially) unknown base with high probability. On the contrary, any CS implementation using only a single or a very small number of pre-stored sensing matrices loses its generality to perform well independent of the signal source. To alleviate this issue, our approach therefore tries to minimize all related storage and memory costs to support a large number of different random sensing matrices, which can be dynamically generated at runtime.

3   Sub-V_T CS Processor

Resource constrained environments, such as ULP processing nodes, pose significant challenges for the implementation of the presented data compression algorithm (Algorithm 1). The key performance issue lies in the realization of the random number sequences needed to address the elements in the sensing buffer. Hence, the goal of our custom designed ASIP architecture is to provide support for an efficient random number sequence generation, enabling energy efficient operation in the sub-V_T regime.

3.1   Processor Baseline Architecture

In this study we use a custom 16-bit reduced instruction set computing (RISC) architecture (TamaRISC [13]), as shown in Figure 1, as the baseline microprocessor. TamaRISC provides a complete RISC instruction set, a C-Compiler, as well as interrupt capability for basic embedded real-time operating system support.

Core Architecture. The main focuses of the architecture lies on minimizing the instruction set complexity in a true RISC fashion, while still providing enough hardware support, especially regarding addressing modes, for efficient execution of signal processing applications.

The microarchitecture has a 3-stage pipeline, comprised of a fetch, decode and execute stage. The core operates on a data word width of 16-bit, comprises 16 general purpose working registers and 3 external memory ports, one for instruction fetch, one for data read and one for data write. The register file has 3 read ports and 4 write ports, and provides 32-bit double word writeback support. Instruction words are 24 bit wide, with every instruction using only a single word. All instructions generally execute in one cycle, which is guaranteed by the use of complete data bypassing inside the core for register as well as memory writeback data.

Moreover, the TamaRISC architecture supports memory-to-memory arithmetic instructions with advanced operand addressing modes, and is hence not a typical load/store RISC architecture, but rather inspired by typical microcontroller architectures.

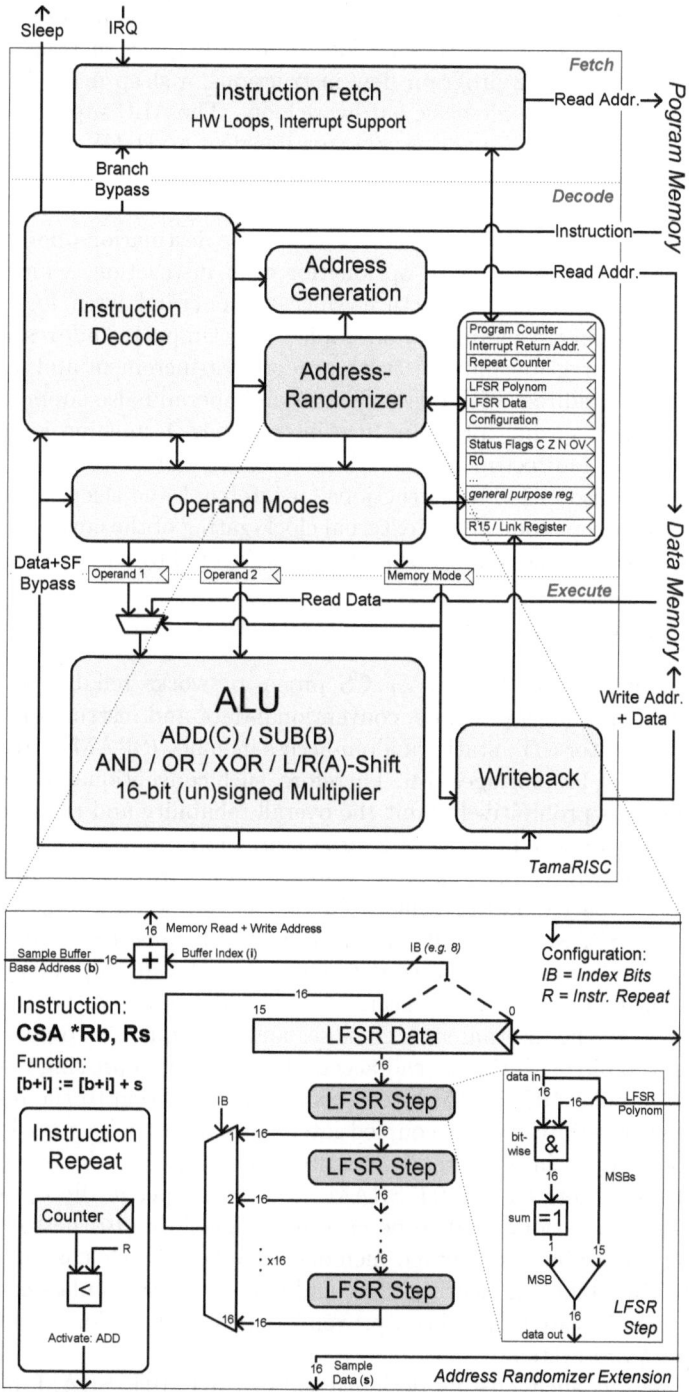

Fig. 1. TamaRISC sub-V$_T$ microprocessor architecture including address-randomizer extension for CS

Instruction Set. The instruction set architecture (ISA) comprises a total of 14 unique instructions, with 8 arithmetic logic unit (ALU) instructions, 2 general data move instructions, 2 program flow instructions, a sleep mode instruction, and an instruction to provide basic hardware loops. The ALU supports addition, subtraction (each with optional carry/borrow), logical AND, OR and XOR, right (arithmetic or not) and left shift, as well as full 16-bit by 16-bit multiplication (32 bit-result) on unsigned and signed data.

All ALU instructions work on two source and one destination operands, using the exact same addressing mode options for each instruction, which helps to reduce complexity of the architecture, since the operand fetch logic and the arithmetic operation are completely decoupled. The supported addressing modes are register direct, register indirect (with pre- or post-increment and decrement) as well as register indirect with offset. The second operand also supports the use of 4-bit literals. Regarding program flow instructions, branching is possible in direct and register indirect mode, as well as by offset with 15 different condition modes. The ISA also includes instructions for interrupt and sleep mode support of the core. The sleep mode allows external clock-gating of the entire core, until a wakeup event occurs (e.g., an interrupt request triggered by a new ADC sample).

3.2 Sub-V_T Memories

While the core logic of the sub-V_T CS processor works reliably at low supply voltages in the sub-V_T regime, conventional data and instruction memories based on 6-transistor (6T) static random-access memory (SRAM) bitcells fail to operate reliably at low voltages [14]. Therefore, such conventional, embedded 6T SRAM macrocells prohibitively limit the overall reliability and the manufacturing yield of the proposed sub-V_T CS processor. More precisely, under gradual supply voltage down-scaling, read and write access failures start to appear first, before the occurrence of data retention failures at even lower voltages [15]. Specially designed SRAM macrocells based on 8- or 10-transistor (8-10T) bitcells are typically used to enable reliable data storage in the sub-V_T regime. For example, a typical 8T SRAM cell contains a read buffer to avoid the direct access of the bit lines to the internal storage nodes and consequently to avoid the risk of switching the bitcell during a read access [16], thereby improving read-ability. Moreover, a popular 10T SRAM bitcell contains, in addition to the read buffer, a tri-state inverter in the cross-coupled latch; this tri-state inverter is disabled during a write access in order to avoid write contention [17], thereby improving write-ability. All these 8T or 10T SRAM macrocells, specifically optimized for robust sub-V_T operation, need to be custom-designed due to the lack of good, commercially available low-voltage memory compilers. Such custom design is associated with a high engineering effort and bares high risk, unless each macrocell is first manufactured and silicon-proven independently, before its integration into a larger VLSI system.

As opposed to such custom-designed sub-V_T 8T/10T SRAM macrocells, we employ a fully automated standard-cell based memory (SCM) compilation flow [18]. The use of SCMs considerably simplifies the design process, and the

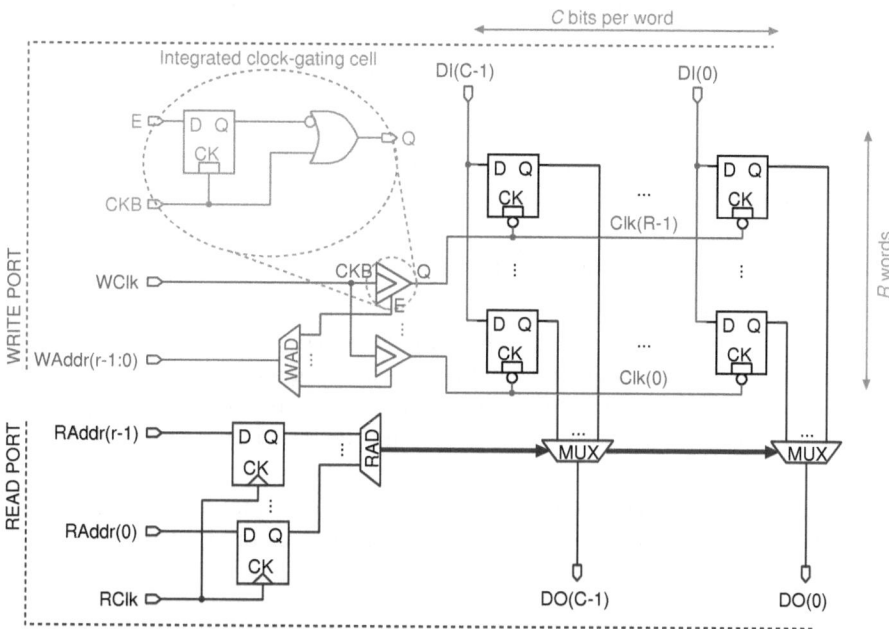

Fig. 2. Schematic of the latch array with clock-gates for the generation of write select signals and static CMOS readout multiplexers. The write port is highlighted in red, while the read port is highlighted in blue.

resulting latch or flip-flop arrays directly avoid the aforementioned reliability concerns of conventional 6T SRAM. In particular, standard-cell latches already contain a read buffer to avoid read failures and a cell-internal keeper which is disabled during write, i.e., during the transparent phase of the latch, to avoid the risk of write contention and write failures. Consequently, the proposed SCMs work reliably in the sub-V_T regime without the need for any extra engineering effort, and allow the complete system to operate at aggressively scaled voltages. Among many architectural variants summarized in [18], this work adopts the latch array architecture shown in Fig. 2. This architecture consists of a write address decoder (WAD) and clock gates for the generation of the one-hot encoded write select pulses (row-wise gated clock signals). Moreover, static CMOS multiplexers are used to read out the desired address (word). While a read logic based on tri-state buffers exhibits lower leakage current, the chosen CMOS multiplexers are faster and more robust for sub-V_T operation. This latch array architecture can be synthesized from commercially available standard-cell libraries. However, note that it is possible to customize one or several standard-cells to meet a specific design goal, such as ultra-low leakage power. For example, the leakage power and access energy can be reduced by approximately 50% by using a single custom-designed standard-cell, namely an ultra-low leakage latch using stack forcing and channel length stretching, as well as a tri-state enabled output buffer to implement the read logic [19].

Even though these latch-based memories are optimized for low-voltage and low-power operation, they still consume considerable leakage power. In our system example, memories account for 70–95% of the architecture's total power consumption, depending on the mode of operation. Furthermore, the sub-V_T memories consume a considerable area share: our implementation with moderate memory sizes of 256 instructions (6 kBit) and 512 data words (8 kBit) results in the processing core only consuming 16% of the total area.

3.3   Index Sequence Implementations

As discussed before, the generation of random numbers used as the buffer indices in Algorithm 1 is commonly performed using one of the following two approaches. The first approach employs precomputation and storage of all required indices in form of a large array in data memory, while the second approach performs the computation of the index sequence at runtime based on a pseudo RNG.

Precomputation. The storage of a preconstructed sequence effectively trades computational effort for memory consumption. For example, the requirement for a single sensing matrix (with 12 non-zero entries per column), used for the compression of a set of 512 samples by 50%, is 6 Kbyte of memory. However, a relatively large memory footprint is especially undesirable in an ULP embedded system, for reasons of die area and power consumption. Since sub-V_T memories are large and consume most of the total power through leakage for low voltages, the storage of tens of Kbyte of data for sensing matrices is not a feasible option, especially when different matrices are to be supported.

Computation at Runtime. The generation of suitable random indices can also be performed by sequence computation based on pseudo RNGs, such as the algorithm proposed in Section 2.2. This approach only requires the data memory to comprise the sensing buffer, which for compression of a set of 512 samples by 50% equals 256 data words (e.g., 512 bytes with a sample precision of 12 (up to 16) bit). As shown in Section 2.2, for each generated index the RNG has to perform the same number of LFSR steps as the number of bits per index. A typical implementation (on a RISC ISA) in software can perform one 16-bit LFSR step in about 10 operations. For the example of a sensing buffer size of 256 and 12 ones per column in the sensing matrix, this results in 12×8 steps per sample, i.e., a computational requirement of about 1 kOp per sample, dedicated to the task of random number generation alone. This requirement becomes problematic, since downscaling of the supply voltage considerably limits the maximum core clock frequency (cf. Figure 3). Due to the relatively large computational overhead, achievable sampling rates for sub-V_T operation are therefore reduced to the range of tens of Hz, which is undesirable.

To combine the benefits of instant random number access of the storage approach, with the memory savings of the computational approach, we propose an instruction set extension for TamaRISC, which performs the task of pseudo-random index generation efficiently in hardware.

3.4   Instruction Set Extension for CS

Analysis of the CS kernel loop shows that the extension of memory operand addressing with efficient randomization can result in significant performance gains. We hence introduce an extension to the TamaRISC instruction set architecture, adding a new instruction that performs an accumulation of sample data on randomized memory addresses within a defined buffer. Essentially, lines 3-6 of Algorithm 1 are combined into a single instruction, named Compressed Sensing Accumulation (CSA). The assembler semantic of CSA is: *CSA *Rb, Rs*. As shown in Fig. 1, the CSA instruction takes two general purpose registers as arguments: the first (Rb) holding the data memory base address (b) of the sensing buffer, the second (Rs) containing the sample data (s). The CSA instruction addresses a random element (i) within the referenced buffer and adds the provided sample onto the existing value in the memory. This operation is repeated for a configured number of iterations, by the use of a counter register dedicated to the instruction. With each repetition a new pseudo random element of the buffer is addressed.

Since the LFSR state of the *address randomizer* can be directly accessed through the register file, the LFSR hardware can also be used for efficient pseudo random number generation, independently of the CS specific memory addressing and accumulation.

Moreover, the CSA instruction is generally used in a small loop in conjunction with the processor's sleep mode, which puts the core in a dormant state (clock-gated) to significantly reduce power consumption until new sample data is available. On wakeup by an interrupt request, the sample data is fetched from the ADC and the CSA instruction is executed, after which the core can immediately be put to sleep again.

Configurability. To enable the construction of many different sensing matrices, the custom instruction is based on four parameters, accessible through dedicated configuration registers. The custom instruction supports software reconfigurability regarding the employed 16-bit LFSR polynomial, the LFSR seed, and the required index width used for memory addressing. Additionally the number of non-zero entries per matrix column can be configured, which equals the number of times a sample is added to pseudo random locations of the sensing buffer. This configurability amounts to storage requirements of at most three 16-bit values per sensing matrix.

Hardware Implementation. The internal hardware structure of the *address randomizer* extension to the TamaRISC micro-architecture is presented in Fig. 1. The custom instruction employs for the sample accumulation the existing 16-bit adder unit in the ALU and does not introduce any new units to the data path of the execution stage of the processor. The decode stage holds the extended address generation logic, which enables addressing of a random word inside the sensing buffer by combining a buffer base address (b) with index bits (i) taken from the least significant bits of the current LFSR state. The number of index bits depends on the value set in the configuration register. In one cycle, the

LFSR state is updated by the same number of LFSR steps as index bits used (1-16). Additionally, the instruction set extension (ISE) is realized as a multi-cycle instruction, which allows handling of one sample in a number of cycles equal to the configured number of non-zero entries (I) per matrix column.

4   Power and Performance Results

Due to the need to retain their memory content, many sensing platforms can not be power gated completely [5], and hence, leakage power is always dissipated. Therefore, our sub-V_T CS processor always operates at a clock frequency that barely accomplishes the task on time while lowering the supply voltage to the corresponding minimum possible level that avoids timing violations. Note that the objective is to minimize power for a given workload in contrast to the operation at the energy-minimum-voltage (EMV), where maximizing energy efficiency often requires a higher operating voltage to balance leakage and active energy.

4.1   Synthesis and Energy Profiling

The design is synthesized above threshold at nominal supply voltage of $1.0\,\mathrm{V}$ with a low-power high threshold-voltage 65-nm CMOS technology, which has a threshold voltage $V_T \leq 700\,\mathrm{mV}$. Toggling information is obtained by simulating a fully routed design (including clock tree) with back-annotated timing information. The design is characterized by employing the sub-V_T energy characterization model that has been derived in [20] and that is briefly introduced in the next subsection. With this model, parameters retrieved from critical path information as well as a traditional *value change dump* based power simulation are used to compute maximum operational speed, energy and power dissipation in the sub-V_T region.

In our implementation, the post-layout critical path delay at nominal supply voltage is $5.2\,\mathrm{ns}$, according to the gate-level static timing analysis. Optimization for maximum frequency and thus a larger slack on the critical path allows for a more aggressive voltage scaling. However, leakage and active power increase considerably with hard timing-constrained designs. Tight constraints will force the tool to infer nets with high fan-out as well as stronger buffers, which increases capacitance on the critical path and consequently yield a slower operation in the sub-V_T region. Following the strategy proposed in [21], we relax the timing constraint to achieve a design with low area and leakage cost. Simulation results show that a relaxed timing constraint of $9\,\mathrm{ns}$, at nominal supply voltage, gives good power results, while it still enables for aggressive voltage scaling for our target applications.

4.2   Sub-V_T Energy Profiling

The total energy dissipation E_T of static CMOS circuits operated in the sub-V_T regime is modelled as

$$E_T = \underbrace{\alpha C_{\mathrm{tot}} V_{\mathrm{DD}}^2}_{E_{\mathrm{dyn}}} + \underbrace{I_{\mathrm{leak}} V_{\mathrm{DD}} T_{\mathrm{clk}}}_{E_{\mathrm{leak}}} + \underbrace{I_{\mathrm{peak}} t_{\mathrm{sc}} V_{\mathrm{DD}}}_{E_{\mathrm{sc}}}, \qquad (2)$$

where E_{dyn}, E_{leak}, and E_{sc} are the dynamic, leakage, and short-circuit energy, respectively. The energy dissipation E_{sc} has been shown to be negligible in the sub-V_T regime [22]. The switching current causing the energy dissipation E_{dyn} results from subthreshold currents [23], i.e., from the drain currents of MOS transistors whose gate-to-source voltage V_{GS} is equal to or lower than the threshold voltage V_T ($V_{GS} \leq V_T$). Whenever the subthreshold current is not used to switch a circuit node, it contributes to E_{leak}. For a given clock period T_{clk}, (2) may be rewritten as

$$E_T = \mu_e C_{inv} k_{cap} V_{DD}^2 + k_{leak} I_0 V_{DD} T_{clk}, \tag{3}$$

where I_0 and C_{inv} represent average leakage current and input capacitance of a single inverter, respectively. Furthermore, k_{leak} and k_{cap} are the average leakage and capacitance of the circuit, respectively, both normalized to a single inverter. Moreover, μ_e represents the circuit's average switching activity.

In the sub-V_T domain, it is beneficial to operate at the maximum achievable frequency to reach minimum energy dissipation per operation. The critical path delay of the circuit is given by

$$T_{clk} = k_{crit} \frac{C_{inv} V_{DD}}{I_0 e^{V_{DD}/(nV_T)}}, \tag{4}$$

where k_{crit} is the critical path delay of the circuit normalized to an inverter delay, $V_T = kT/q$ is the thermal voltage and n is the subthreshold factor. By assuming that operation is performed at the maximum frequency, the total energy dissipation E_T is found by introducing (4) into (3), which gives

$$E_T = C_{inv} V_{DD}^2 \left[\mu_e k_{cap} + k_{crit} k_{leak} e^{-V_{DD}/(nV_T)} \right]. \tag{5}$$

Additionally, for a system that operates at a fixed frequency with a given clock period T_{clk}, at a given V_{DD}, the power is derived from (3) as

$$P = \frac{\mu_e C_{inv} k_{cap} V_{DD}^2}{T_{clk}} + k_{leak} I_0 V_{DD}, \tag{6}$$

which shows that the static power consumption is directly proportional to V_{DD}.

The application of the model provides the sub-V_T profile of a design, i.e., energy/operation, power consumption, and critical path speed. The model was validated by measurements and accuracy is within a 10% error rate at the measured temperatures 0, 27 and 37 °C. In Table 1 the design properties of the sub-V_T processor are shown in terms of leakage, capacitance and critical path normalized to a single inverter, as well as switching activity. For more details, the reader is referred to [20].

4.3   Simulation Results

Fig. 3 shows the power consumption and the corresponding supply voltage of the sub-V_T CS processor for various clock frequencies in the sub-V_T domain,

Table 1. Architectural properties for sub-V_T modeling

Design properties	Value
k_{cap}	254 000
k_{crit}	434
k_{leak}	194 000
μ_e	0.0675

computed using (6) and (4). More specifically, a clock frequency of 100 kHz
for the CS processor is achieved at a supply voltage of 0.37 V. As a result, a
total power of 288 nW is dissipated, where 27% of the power consumption is
due to leakage power. When the required clock frequency is reduced to 1 kHz
through voltage scaling, the sub-V_T CS processor consumes 22.5 nW in total,
where the leakage dissipation now has a share of 98%. As demonstrated in Fig. 3,
the leakage power dominates the overall power for clock frequencies lower than
1.5 kHz, corresponding to 0.2 V supply voltage. In this particular technology,
operation below 0.25 V is not recommended due to higher rates of functional
failures from larger process variations according to [20].

The energy profile of the sub-V_T CS processor is shown in Fig. 4. Energy
dissipation at maximum operational frequency (5) is shown together with fixed
clock frequencies (3) of 2.1 kHz, 16.5 kHz and 100 kHz. It is observed that
operating at lower frequency than dictated by supply voltage results in higher

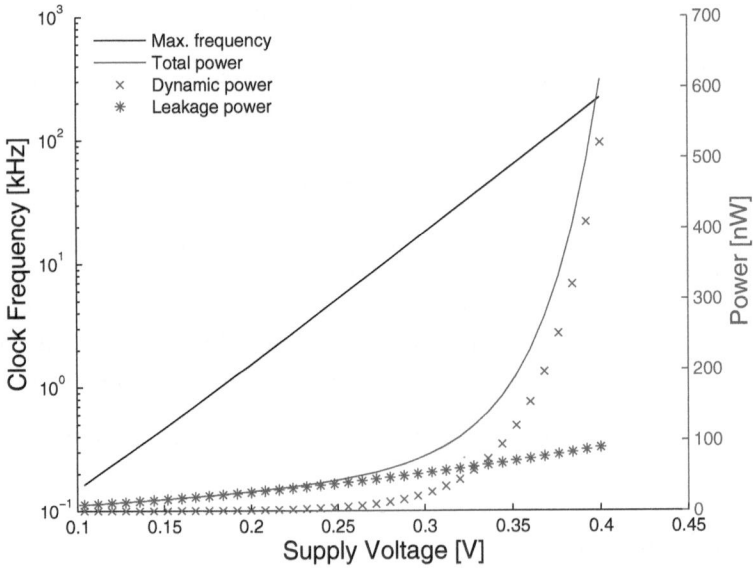

Fig. 3. Power and performance exploration of the sub-V_T CS processor

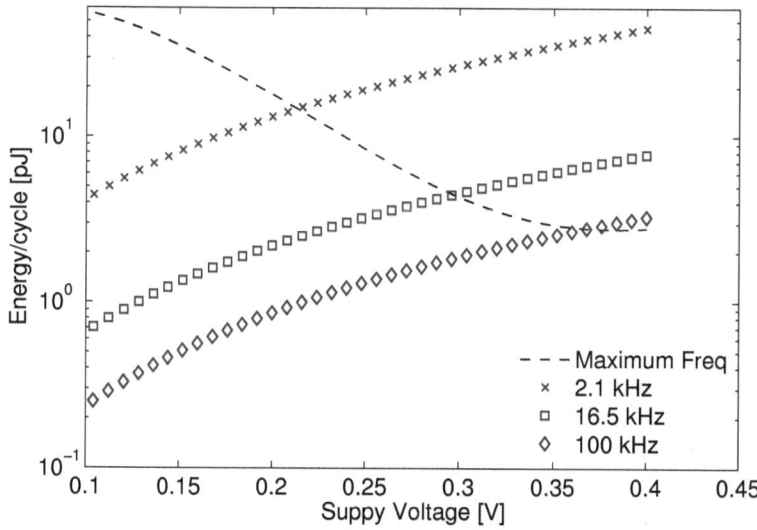

Fig. 4. Energy profile for various operational frequencies of the sub-V$_T$ CS processor

energy dissipation. Thus the implementation goal is to use a supply voltage that is just barely sufficient to support the necessary clock frequency.

The total area of the sub-V$_T$ CS processor is 84.7 kGEs, where 1 GE corresponds to the area of a NAND-2 minimum drive strength gate. The instruction and data memory in the processor have a size of 768 Bytes and 1 kByte, respectively. The memories occupy 84% of the overall area, whereas the core occupies the rest. The area overhead of the instruction set extension for CS accounts for less than 3% of the overall area.

4.4 Case Study: CS-Based ECG Signal Compression

As a case study, we apply the CS algorithm for the compression of ECG signals [3]. The test case performs data compression on blocks of 512 samples, recorded at different sampling rates.

Quality of Produced Sensing Matrices. Mamaghanian et al. [3] have shown that 12 non-zero elements in each column of the sensing matrix are sufficient to maintain satisfactory quality of reconstructed ECG signals for diagnostic purposes. Based on the study in [3], we group random indices into groups of 12, where each group determines the non-zero elements of the corresponding column in the sensing matrix. Assuming that there are no repeated indices in a group, the corresponding column of the sensing matrix will have only ones and zeros. However, in case of repetition the repeated indices will accumulate, which, according to our experiments, does not lead to any quality degradation in the reconstructed signal as shown in Fig. 5 for an example sensing matrix.

To ensure a good quality of diagnostic analysis on the reconstructed ECG signal, the compression performance is quantified according to the percentage root-mean-square difference (PRD) for different compression ratios [3]. PRD quantifies the percent error between the original and the reconstructed signal where a PRD value less than 9 is classified as "very good" or "good" quality for ECG diagnosis. Thanks to our configurable CS-extension, many sensing matrices with different combination of primitive polynomials and seeds can be constructed. These sensing matrices are analyzed by quantifying their corresponding PRD values for various compression ratios. More specifically, Fig. 5 shows as an example the PRD values with respect to various compression ratios for one of the constructed sensing matrices with a polynomial

$$p = x^{13} + x^{12} + x^{11} + x^{10} + x^9 + x^7 + x^3 + x^2 + x^1 + 1$$

and the seed "0x6218" in hexadecimal combination. As seen from Fig. 5, a PRD value below 10 is retained for compression ratios up to almost 60%. Moreover, 50% compression is achieved with a PRD of 7.7. Similar to the state-of-the-art CS sensing matrices [3], our sensing matrices that are generated by our multi-step LFSR mechanism, accomplish a "good" or "very good" quality of the reconstructed signals for compression ratios less than 53%.

Power vs. Performance Analysis. We consider the example of 50% data compression of ECG signals, using the ECG database in [24] for stimuli generation, to analyze the power and performance of our sub-V_T CS processor. The required operating frequency to support a given sampling rate to compress ECG signals in real-time is given by: $f \geq N * f_s$ where f_s and N stand for the given sampling rate and the required average number of clock cycles to process a sample. The clock frequency of the sub-V_T CS processor is always adjusted, to have the minimum required clock frequency, according to the given ECG sampling rate. The supply voltage of the processor is then lowered accordingly.

The presented sub-V_T CS processor requires 8460 clock cycles to apply 50% compression on 512 samples of ECG data when the sensing matrix is constructed by 12 random indices per column ($I = 12$). This corresponds to only an average of $N = 16.5$ cycles processing time for each sample (16 cycles per sample + setup overhead per sample set). As a result, the sub-V_T CS processor must operate with a clock frequency of 2.1 kHz and 16.5 kHz for 125 Hz and 1 kHz sampling rates, respectively. Fig. 6 shows the power consumption of the sub-V_T CS processor for various ECG sampling rates. More specifically, for 125 Hz sampling rate the sub-V_T CS processor consumes only 30.6 nW in total with 95% of the power due to leakage. Similarly, the total power consumption is only 74 nW for a sampling rate of 1 kHz, where 70.7% is because of leakage dissipation.

To compare our ISE-enhanced CS processor with the baseline processor, we consider the construction of the CS sensing matrix by computing random sequences of indices based on a pseudo RNG algorithm (c.f. Section 2) running on the baseline ISA. Our results show that the optimized implementation for the baseline core requires a significantly higher computational effort. Specifically,

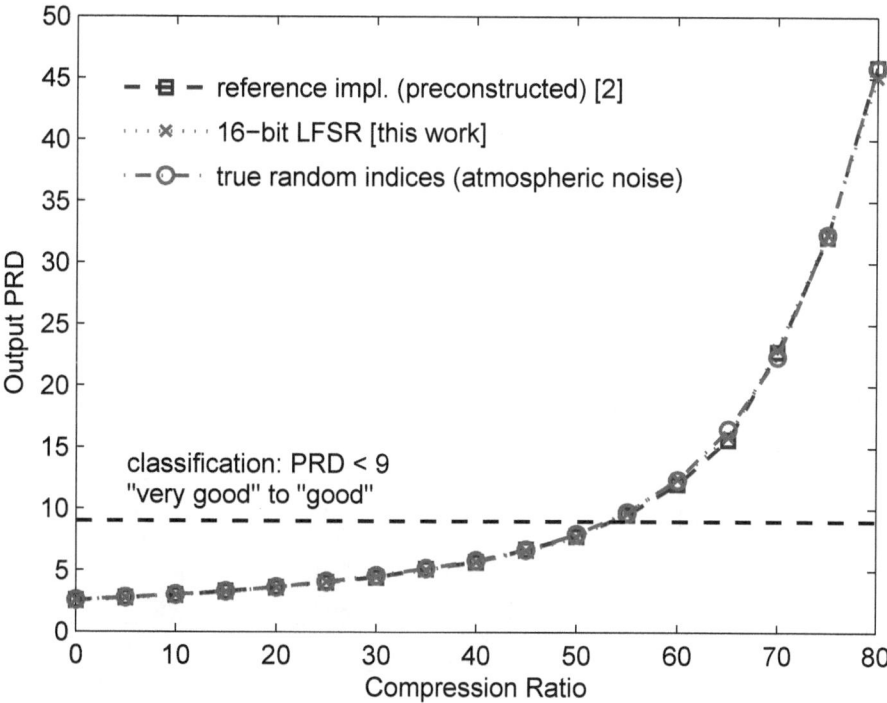

Fig. 5. PRD values at various compression ratios for three index sequences (sensing matrices Φ), each using different methods of construction

the increased computational effort per sample in terms of cycles amounts to $(10log_2(k) + 5)I + 5$, compared to our implementation based on the proposed CSA instruction with an effort of $I + 4$ cycles. In this case of LFSR emulation by software, code optimized to the baseline ISA processes one ECG sample, including the sensing matrix construction, on average in $N = 1025.5$ cycles, which translates into a speed-up of 62× for our ISE-supported implementation. Therefore, a sampling rate of $f_s = 125\,\text{Hz}$ requires a clock frequency of 128 kHz, using the pure software approach. This results in a total power consumption for the design of 355 nW (cf. Fig. 3), which is 11.6× higher than the sub-V$_T$ CS processor with ISE, where the random indices are produced with the help of the embedded LFSR. Moreover, Mamaghanian et al. [3] report a code execution time of 25 ms on a different architecture with a clock frequency of 8 MHz, for applying 51% compression on a set of 512 ECG samples, where pre-computed random indices are stored in the memory. This results in $N = 390.5$ cycles per sample, a 23.6× higher performance requirement than our CS implementation, in terms of cycle count alone.

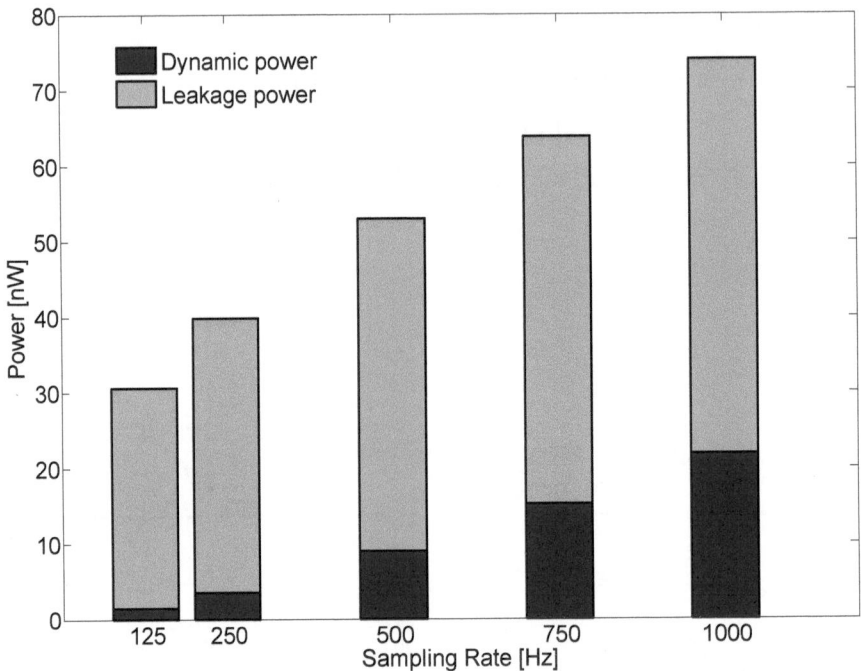

Fig. 6. Power consumption for various ECG sampling rates

5   Conclusion

Compressed sensing (CS) is a well-known universal data compression technique applied to sparse signals, used widely for sensing environment applications. Autonomous and portable devices, such as sensing platforms, however enforce ultra-low-power CS implementations, due to their limited energy resources. Therefore, we have proposed a subthreshold processing platform specifically optimized for CS, while still maintaining the flexibility and configurability of a processor based system. To this end, we have customized the instruction set architecture of a low-power baseline processor to exploit the specific operations of the CS algorithm. Specifically, we propose a Compressed Sensing Accumulation (CSA) instruction that efficiently performs accumulation of sample data on randomized memory addresses within a defined sampling buffer. Moreover, our processing platform embeds the required data and instruction memories in the form of sub-V_T-capable standard-cell memories, which are essential for ULP operation. We show that our processing platform requires neither high computational effort nor excessive memory sizes compared to straight-forward implementations. Therefore, the platform is well suited to exploit subthreshold computing at low voltages and with very low leakage. Our system consumes only 30.6 nW for a case study of CS-based electrocardiogram (ECG) signal compression at an ECG sampling rate of 125 Hz. Our results show that the proposed processing platform achieves 62× speed-up and 11.6× power savings with respect to an established CS implementation running on the baseline low-power processor.

Acknowledgment. This work has been partially funded by the BodyPowered-SenSE Nano-Tera.ch RTD project (ref. number: 20NA21_143069), which is evaluated by the Swiss NSF and funded by Nano-Tera.ch with Swiss Confederation financing.

References

1. Constantin, J., Dogan, A., Andersson, O., Meinerzhagen, P., Rodrigues, J., Atienza, D., Burg, A.: TamaRISC-CS: An ultra-low-power application-specific processor for compressed sensing. In: 2012 IEEE/IFIP 20th International Conference on VLSI and System-on-Chip (VLSI-SoC), pp. 159–164 (2012)
2. Donoho, D.L.: Compressed sensing. IEEE Trans. on Information Theory 52(4), 1289–1306 (2006)
3. Mamaghanian, H., et al.: Compressed sensing for real-time energy-efficient ECG compression on wireless body sensor nodes. IEEE TBME 58(9), 2456–2466 (2011)
4. Jocke, S.C., et al.: A 2.6-uw sub-threshold mixed-signal ecg soc. In: Symposium on VLSI Circuits, pp. 60–61 (2009)
5. Hanson, S., et al.: A low-voltage processor for sensing applications with picowatt standby mode. IEEE J. Solid-State Circuits 44(4), 1145–1155 (2009)
6. Kwong, J., et al.: A 65nm sub-vt microcontroller with integrated sram and switched-capacitor dc-dc converter. In: ISSCC, pp. 318–616 (2008)
7. Dogan, A.Y., Atienza, D., Burg, A., Loi, I., Benini, L.: Power/Performance exploration of single-core and multi-core processor approaches for biomedical signal processing. In: Ayala, J.L., García-Cámara, B., Prieto, M., Ruggiero, M., Sicard, G. (eds.) PATMOS 2011. LNCS, vol. 6951, pp. 102–111. Springer, Heidelberg (2011)
8. Hanson, S., et al.: Exploring variability and performance in a sub-200-mV processor. IEEE J. Solid-State Circuits 43(4), 881–891 (2008)
9. Zhai, B., et al.: A 2.60pJ/Inst subthreshold sensor processor for optimal energy efficiency. In: IEEE VLSI, pp. 154–155 (2006)
10. Dreslinski, R., et al.: Near-threshold computing: Reclaiming Moore's law through energy efficient integrated circuits. Proc. IEEE 98(2), 253–266 (2010)
11. Banz, C., et al.: Instruction set extension for high throughput disparity estimation in stereo image processing. In: ASAP, pp. 169–175 (September 2011)
12. Kwong, J., Chandrakasan, A.: An energy-efficient biomedical signal processing platform. IEEE J. Solid-State Circuits 46(7), 1742–1753 (2011)
13. Dogan, A.Y., et al.: Multi-core architecture design for ultra-low-power wearable health monitoring systems. In: DATE (2012)
14. Qazi, M., Sinangil, M., Chandrakasan, A.: Challenges and directions for low-voltage SRAM. IEEE Design and Test of Computers 28(1), 32–43 (2011)
15. Mukhopadhyay, S., Mahmoodi, H., Roy, K.: Modeling of failure probability and statistical design of SRAM array for yield enhancement in nanoscaled CMOS. IEEE Transactions on Computer-Aided Design of Integrated Circuits and Systems 24(12), 1859–1880

16. Chang, L., Fried, D., Hergenrother, J., Sleight, J., Dennard, R., Montoye, R., Sekaric, L., McNab, S., Topol, A., Adams, C., Guarini, K., Haensch, W.: Stable SRAM cell design for the 32 nm node and beyond. In: 2005 Symposium on VLSI Technology. Digest of Technical Papers pp. 128–129 (June 2005)
17. Jain, S., Khare, S., Yada, S., Ambili, V., Salihundam, P., Ramani, S., Muthukumar, S., Srinivasan, M., Kumar, A., Gb, S., Ramanarayanan, R., Erraguntla, V., Howard, J., Vangal, S., Dighe, S., Ruhl, G., Aseron, P., Wilson, H., Borkar, N., De, V., Borkar, S.: A 280mV-to-1.2V wide-operating-range IA-32 processor in 32nm CMOS. In: 2012 IEEE International Solid-State Circuits Conference Digest of Technical Papers (ISSCC), pp. 66–68 (February 2012)
18. Meinerzhagen, P., et al.: Benchmarking of standard-cell based memories in the sub-V_t domain in 65-nm CMOS technology. JETCAS 1(2), 173–182 (2011)
19. Meinerzhagen, P., Andersson, O., Mohammadi, B., Sherazi, Y., Burg, A., Rodrigues, J.: A 500 fW/bit 14 fJ/bit-access 4kb standard-cell based sub-VT memory in 65nm CMOS. In: Proc. IEEE ESSCIRC, pp. 321–324 (September)
20. Akgun, O., et al.: High-level energy estimation in the sub-V_T domain: Simulation and measurement of a cardiac event detector. IEEE TBCAS 6(1), 15–27 (2012)
21. Meinerzhagen, P., et al.: Synthesis strategies for sub-V_t systems. In: ECCTD, pp. 552–555 (2011)
22. Vittoz, E.: Low-Power Electronics Design. CRC Press (2004)
23. Soeleman, H., Roy, K., Paul, B.: Robust subthreshold logic for ultra-low power operation. IEEE T-VLSI Systems 9(1), 90–99 (2001)
24. Harvard-MIT Division of Health Sciences and Technology Biomedical Engineering Center: MIT-BIH arrhythmia database directory, http://www.physionet.org/physiobank/database/mitdb

Configurable Low-Latency Interconnect
for Multi-core Clusters

Giulia Beanato[1], Igor Loi[2],
Giovanni De Micheli[1], Yusuf Leblebici[1], and Luca Benini[2]

[1] EPFL, Lausanne, Switzerland
[2] DEIS, University of Bologna, Bologna, Italy
{giulia.beanato,giovanni.demicheli,yusuf.leblebici}@epfl.ch,
{igor.loi,luca.benini}@unibo.it

Abstract. Shared L1 memories are of interest for tightly-coupled processor clusters in programmable accelerators as they provide a convenient shared memory abstraction while avoiding cache coherence overheads. The performance of a shared-L1 memory critically depends on the architecture of the low-latency interconnect between processors and memory banks, which needs to provide ultra-fast access to the largest possible L1 working set. The advent of 3D technology provides new opportunities to improve the interconnect delay and the form factor. In this chapter we propose a network architecture, 3D-LIN, based on 3D integration technology. The network can be configured based on user specifications and technology constraints to provide fast access to L1 memories on multiple stacked dies. The extracted results from the physical synthesis of 3D-LIN permit to explore trade-offs between memory size and network latency from a planar design to multiple memory layers stacked on top of logic, evaluating the improvement in both form factor and latency.

Keywords: 3D integration, multi-core processor, shared memory, interconnection network.

1 Introduction

Following Moore's law, the scaling to nanometer technologies has led to a transition from single-core to multi-core processors, and is now moving towards many-cores architectures [1]. Whereas hundreds of millions of transistors can now be placed on a single chip leading to increased computing power, they cannot be fully exploited due to interconnect latency. In nanometer-scale technologies, interconnect latency and power do not scale as much as device geometries, thus becoming a performance bottleneck. These limiting factors need to be overcome at the architectural level. For many applications, the exploitation of customized accelerators will be the way to obtain the highest performance, together with more efficient types of interconnect and memory hierarchies [2].

For this reason, new interconnect architectures have already been envisaged. For instance, *Network-on-chip (NoC)* [3] has been adopted to substitute conventional bus-based systems when high bandwidth and high speed are required.

A. Burg et al. (Eds.): VLSI-SoC 2012, IFIP AICT 418, pp. 107–124, 2013.
© IFIP International Federation for Information Processing 2013

When ultra-low latency processor to memory interconnection is requested for parallel computing, novel fast interconnect topologies are imperative to guarantee the access to the memory in few clock cycles. Several research efforts are already focused on low-latency, high-bandwidth connection between the processing elements and multi-banked on-chip memories. The *Mesh-of-Trees (MoT)* Interconnection Network proposed in [4], the Hyper-core architecture [5] and the single-cycle interconnection network presented in [6] are just few examples of low-latency networks. Nevertheless, future generations of *Chip Multi-Processor (CMP)* require a major innovation in both integration technology and on-chip communication infrastructure.

A promising option to overcome the barrier in interconnect scaling is the 3D integration of integrated circuits (3D ICs)[7]. Stacking multiple chips and connecting them by *Through Silicon Vias (TSVs)* has the potential to reduce the interconnect wire length while offering high vertical connect density. Multi-cores and many-cores processors can benefit from several characteristics of 3D devices: (a) Wire length reduction improves the latency of core to memory interconnect; (b) High TSV density and their small length can be exploited for improving memory bandwidth when stacking memory layers on top of logic layers; (c) The smaller form factor due to the addition of a third dimension is essential for moving on-chip the memory required by the processing elements avoiding slow off-chip connections.

In the last few years, several studies have been published exploring 3D integration technology in order to address the high area overhead of SRAM. A proposal from Li et al.[8], focuses on the L2 cache design and management in a 3D chip. They propose a network architecture embedded into the L2 NUCA cache memory for connecting it to a collection of cores. A different approach is followed by Loh, that in [9] considers 3D-DRAM stacked on top of multi-processors and revises the memory system organization in a 3D context. More recently, also Woo et al.[10], have explored a memory architecture that exploits TSVs for connecting the last level cache to the 3D stacked DRAM. The work of Madan et al.[11] instead, takes in consideration a 3D system composed by a DRAM layer and an SRAM cache banks layer on top of a processing layer. Considering emerging memory technologies, Mishra et al.[12] study the integration of STT-RAM in a multi-core system, together with a network level solution for decreasing the write latency associated with these novel memories.

In order to connect memory and logic placed on different layers, several groups already explored a methodology to extend NoC design into a 3D setting. The simple extension of traditional NoC fabrics to the third dimension adding routers at each layer (Symmetric NoC), does not pay in performance due to the different delay between fast vertical TSV and the horizontal interconnects. A first proposal has been done by Li et al. [8], with a network architecture embedded into the L2 cache memory. The use of Time-Division Multiple Access (dTDMA) buses as Communication Pillars between the wafers is proposed in order to have single-hop communication amongst the layers. The 3D Dimensionally-Decomposed(DimDe) Router [14], focus on optimizing of the inter-strata communication with single hop connection between any two layers. Park et al. [15] propose a Multi-layered on-chip Interconnect Router Architecture (MIRA) divides the NoC between the multiple layers optimizing the micro-architecture for Non Uniform Cache Architecture (NUCA)-based CMP. A Low-Radix Low-Diameter 3D Interconnection Network is proposed by Xu et al. [16] which adopts long wires to connect remote intra-layer nodes and results in a 3 hops diameter network. More recently, Xue et al. [17] uses long range links to replace multiple short links in order to build a 5 hops 3D interconnection network for many core processors that exploits the DimDe router. While Ben Ahmed et al.[18] focus on overcoming the limitations in power, communication cost and throughput of their 2D OASIS-NoC by extending it to 3D.

This chapter aims to propose a fully synthesizable *3D Logarithmic Interconnection Network (3D-LIN)* for connecting a cluster of processing elements, placed on a logic layer, to multiple layers of SRAM modules. These modules constitute a single shared L1 memory that can enable fast communication among the tightly coupled processing elements avoiding cache coherence overheads. The network is configurable in both 2D and 3D-domains and is automatically split between the chosen number of memory layers. In order to reduce the chip cost, regardless of the number of memory layers needed, they all have the same layout and can all be produced exploiting the same mask. Design automation and configuration of the network allow us to experiment with different 3D structures, in the search for the trade-off points between speed, footprint and number of layers.

2   2D Network

The basic 2D-LIN is a low-latency and flexible crossbar that connects multiple *processing elements (PEs)* to multiple SRAM *memory modules (MMs)*. The IP is designed and optimized for sustaining full bandwidth and supporting non-blocking communication within a single clock cycle. These features makes LIN an interesting option for interfacing multi-processors to a shared Tight Coupled Data Memory (TCDM) constituted by multiple equal memory banks. This topology permits to avoid data replication providing also a simple and fast way for inter-processors communication and multi-core synchronization. In order for the design to be simple and efficient, the interconnect is built following the Mesh Of Trees approach, where the network is created combining binary trees. Each tree provides a unique combinational path between the processing element cluster and one memory module, and viceversa. Aiming to sustain non blocking communication, the request and the response path must be decoupled, hence 2D-LIN features independent request and response network. The key property of this soft IP is the reconfigurability. The user has control on a number of parameters:

- Number of masters, N, that is a power of two;
- Number of memory cuts, M, that is a power of two. With a number of MMs at least double the numbre of PEs, access collision can be drastically reduced;
- Size of the memory cuts, all the banks should have the same size;
- Data and Address width;

- Enable for word level interleaving, for spreading transactions among all banks drastically reducing access collision.
- Test and Set bit. This bit act as enable for a test-and-set instruction used to write to a memory location and return the old value as a single atomic operation.

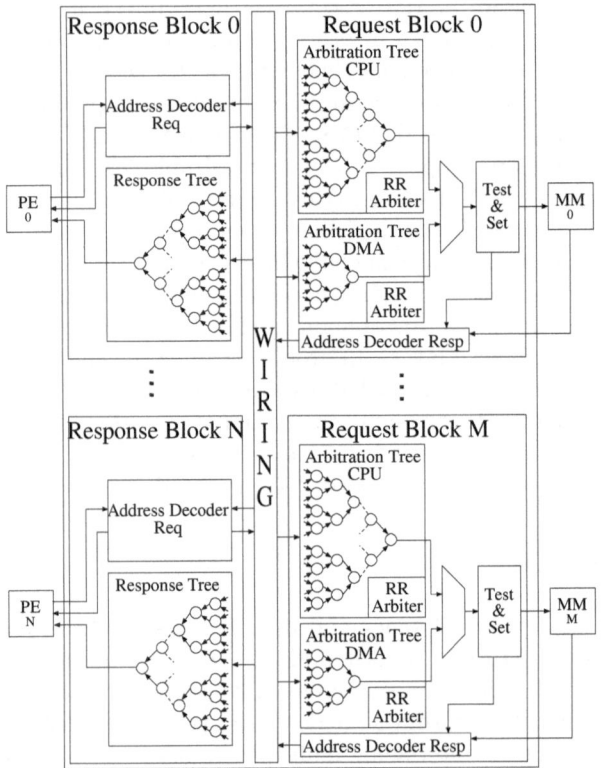

Fig. 1. Block schematic of the 2D-LIN

2.1 Network Architecture Protocol

The network is created by independent and decoupled Request and Response channel. A memory access starts with a request issued by a PE through a master port, then, the master is kept updated on the status of the request by a simple and lean protocol based on a credit based flow control. Each clock cycle, all the requests made from PEs are propagated through the binary trees. Collisions due to multiple requests directed to the same memory bank are avoided by Round Robin arbitration performed at each node. The processors losing the arbitration are stalled. The PE winning the arbitration concludes the transfer in a single clock cycle in case of a store, while, in case of a load, the read data is returned the next clock cycle.

2.2   Request Block

The request block is in charge of collecting all the PE's requests directed to a specific memory module (see Figure 1). In the simplest case of two PEs, the block is built out of a single binary tree where the request block is composed of 1 node, being a routing-arbitration primitive. The number of stages of the Arbitration Tree is a function of the number of masters attached to it: $NUM_{stage}=\log_2(N)$, N being the number of PEs. Combining several binary trees, the network can support both generic number of ports and different priorities. Hence, a high priority channel for the processors and a low priority channel for eventual peripherals can be supported. The primitives composing the request block first arbitrate among eventual requests through a Round Robin policy, then the winning one is routed to the MM in a combinational way. At the same time, the flow control signals traveling from MMs to PEs, are also managed. Both normal read/write operation and atomic test and set are supported.

2.3   Response Block

The response block (see Figure 1) is in charge of collecting all the responses from memory modules which are directed to a specific processing element, therefore, it can be considered as a specular version of the request block. Nevertheless, since the response network is only used for read operations and the read latency is deterministic (1 cycle), no response collisions are possible. Hence, the response path does not need any arbitration, and it can be simplified replacing round robin arbiters with simpler decoders.

3   3D Interconnection Network

Within a standard planar(2D) architecture, when more storage capability or more processing power are needed, the network size increases, and the single-cycle communication becomes the limiting factor for the maximum achievable operating frequency. 3D-LIN is the extension of the 2D structure presented in the previous section, to be integrated in a 3D-stacked CMP. This network topology allows designers to overcome the limitation in frequency by automatically splitting the 2D floorplan into one logic layer and several memory layers and stacking them one on top of the other, Figure 3. All the power-hungry processing elements are placed on logic layer, close to the heat sink, while the memory banks, are divided among the memory layers. The network is partitioned among the layers in an automated way following the assumption that all the memory layers should have the same identical layout:

- Each layer automatically auto-configures during runtime. This permits to reduce the chip cost and the design effort.
- TSVs from the bottom layer are connected to the lowest metal layer, while the TSVs to the upper layer are connected to the top metal layer.

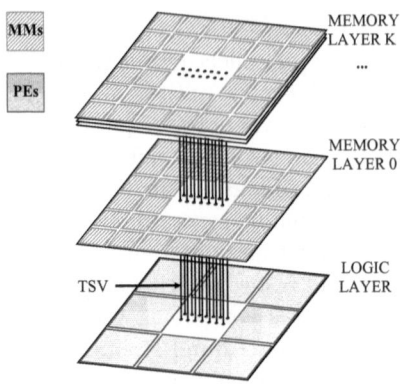

Fig. 2. 3D chip architecture

- The M memory banks are equally divided among K memory layers, where K is a power of 2. Each memory layer contains $M_L=M/K$ memory banks.

Table 1 summarize the main parameters of 3D-LIN versus 2D-LIN. We can notice that in terms of number of levels of the trees, the first strongly depends on the number of PEs, while the second is related to the number of MMs. The number of levels directly affects the latencies of the request network path (PE to MM), and the response path (MM to PE). When connecting the memory banks, the access time to read the data from the memory is added to the latency of the response path. 3D-LIN allows us to decrease the number of arbitration levels of the response tree when implemented on 2 or more memory layers, hence it allows the system to run at higher frequencies. The number of primitives per layer and in the system give an estimation on how the area of the network can be reduced by moving to 3D. The main reduction is encountered for the primitives of the Response Tree, but also the Arbitration Tree diminish.

3.1 Network Architecture

TSVs connecting the stacked dies have good electrical characteristics, but their area footprint is bigger compared to the on-chip metal lines. For this reason it is important to place the minimum number of TSVs, while still guaranteeing the maximum possible bandwidth. When the signals traversing the tiers are the direct input and output of the processor, it is possible to place the minimum number of TSVs dedicated to signal propagation:

$$TSV = (Nc + 1 + log_2 K) + N(Nb_{addr} + 2Nb_{data} + Nb_{byteEN} + 2) \qquad (1)$$

where Nc is the number of TSVs for clock propagation, summed to one TSV for the reset signal, log_2K is the number of bits needed for the layer ID. Nb_{addr}, Nb_{data} and Nb_{byteEN} are respectively the number of TSVs for propagating the

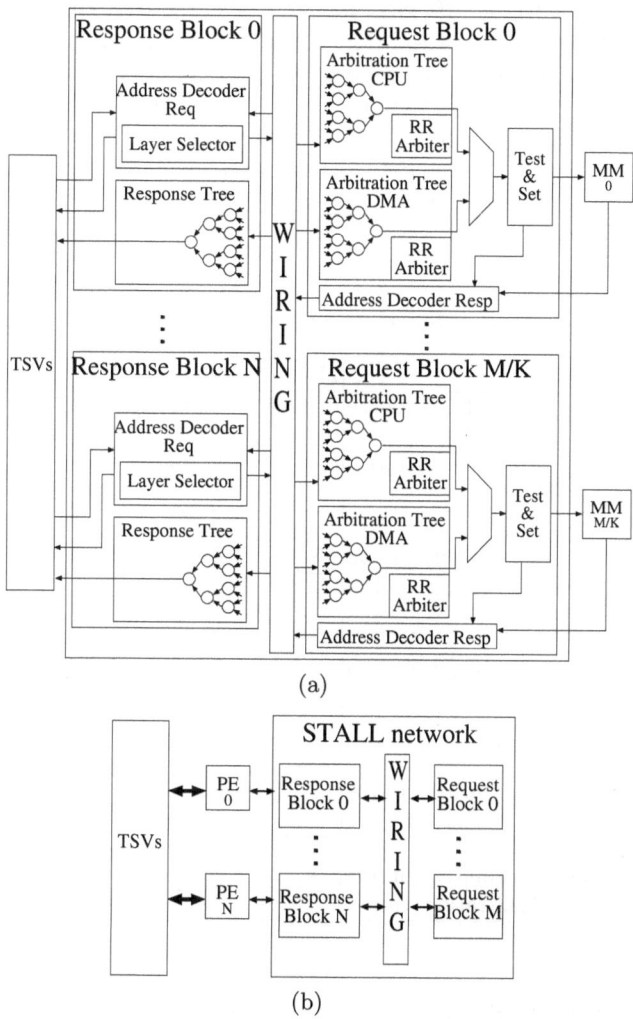

(a)

(b)

Fig. 3. Block schematic of the 3D-LIN: (a) Logic layer block diagram; (b) Single memory layer block diagram

address, the data and the byte enable signals. The maximum bandwidth of the 2D system is:

$$BW_{max} = f(\frac{Nb_{data}}{8})K \tag{2}$$

Hence, the PEs and the small Network for the stall (see Figure 3(b)) are placed on the logic layer, while each memory layer has the same layout and contains a Network of cardinality $N \times \frac{M}{K}$ and $\frac{M}{K}$ memory banks (see Figure 3(a)). This configuration that minimize the number of TSVs needed for the signals, still guarantee BW_{max} also for the 3D implementation. The layerID signal is sent from the logic layer to identify each memory layer, so that the address space is equally divided between all the MMs. Each memory layer takes the incoming layerID as its own identifier, and send to the next mem layer the received signal incremented by one. In the TSV count, the Stall signal is not taken in account. In the 2D network, the Stall signal is critical, because it needs to be asserted much in advance with respect to the next clock rising edge. Hence, in order to optimize it, the logic that computes the Stall signals is detached from the main Network connecting PEs to MMs and placed on the logic layer as a small independent Network.

Table 1. 3D-LIN vs. 2D-LIN

	2D-LIN	**3D-LIN**
Number of levels Response Tree	$log_2 M$	$log_2 \frac{M}{K}$
Number of levels Arbitration Tree	$log_2 N$	$log_2 N$
Number of primitives on each memory layer - Response Tree	$\displaystyle\sum_{i=1}^{log_2 M} \frac{M}{2^i} \times N$	$\displaystyle\sum_{i=1}^{log_2 \frac{M}{K}} \frac{\frac{M}{K}}{2^i} \times N$
Number of primitives on each memory layer - Arbitration Tree	$\displaystyle\sum_{i=1}^{log_2 N} M \times \frac{N}{2^i}$	$\displaystyle\sum_{i=1}^{log_2 N} \frac{M}{K} \times \frac{N}{2^i}$
Number of primitives in the system - Response Tree	$\displaystyle\sum_{i=1}^{log_2 M} \frac{M}{2^i} \times N$	$\displaystyle\sum_{j=1}^{K} \sum_{i=1}^{log_2 \frac{M}{K}} \frac{\frac{M}{K}}{2^i} \times N$
Number of primitives in the system - Arbitration Tree	$\displaystyle\sum_{i=1}^{log_2 N} M \times \frac{N}{2^i}$	$\displaystyle\sum_{j=1}^{K} \sum_{i=1}^{log_2 N} \frac{M}{K} \times \frac{N}{2^i}$

3.2 Network Operation

During a read/write operation, the master asserts data and control signals that are sent as a packet. Some control signals go to the Stall Network that arbitrates possible collision and eventually sends the Stall signal to the PE within the same clock cycle. The full packet, data and control signals, are also sent through the TSVs to the memory layers. Each memory layer receives the packet and checks if the request is for a position in its address range. The layer containing the address lets the packet enter, while the other layers invalidate the request. When a packet accesses the memory layer containing the requested address, the network routes and arbitrates the packet among the other simultaneous requests, allowing the higher priority request to access the memory bank. Write operations are performed in the same clock cycle, while for Read operations and Test and Set operations, the read data is propagated back to the related PE in the next clock cycle.

4   Experimental Results

This section provides the evaluation of 3D-LIN in terms of area, power and delay. The Network is implemented in System-Verilog and synthesized with Synopsys Design Compiler in topographical mode using 65nm CMOS technology library from ST-Microelectronics. The physical synthesis has been chosen to extract the results because it allows the user to floorplan the design and accurately predict post-layout timing using real net capacitances during RTL synthesis [19]. The functionality has been verified using Mentor Graphics' Modelsim.

In this experiment we considered $5\mu m$ wide TSV with $10\mu m$ minimum pitch and a length of $50\mu m$, which represents the state-of-the-art for high density through silicon vias [20]. According to the chosen dimensions, the TSV's parasitic capacitance have been obtained through the analytical model proposed by Kim,[21]. For the experiments, the parasitics values have been rounded to $20m\Omega$ for the resistance and 30fF for the capacitance.

The memory size depends on the multi-core application. For the experiments, we chose a case study with memory modules chosen to be SRAM banks of 8kB, which timing and physical information are provided by the lib file and the Milkyway database. Each MM occupy $0.06mm^2$. Regarding the processing elements, dummy hard macros are used in order to emulate their area occupation. Each PE is considered to be an ARM CortexM3, which the estimated area is around $0.07mm^2$ for 65nm technology.

Unfortunately, the current version of Synopsys DC does not support TSV and 3D stacking, hence, in the absence of established design kits, the synthesis flow is performed in several main steps. Starting from the synthesizable RTL description of the network, already configured with the user constraints, the floorplanning of memory layer is performed, and the time and physical constraints are added to emulate the TSVs. After the physical synthesis of the memory layer, the back-annotated delays are used to perform the physical synthesis of the logic layer. After the floorplan definition, the logic layer is synthesized considering the latencies of the stacked dies. These steps are then iterated to meet the desired timing constraints for the complete 3D-stacked system.

4.1   Physical Analysis

When moving to a 3D configuration, the original NxM network is divided among the layers: a small NxM network for the Stall signal is placed on the logic layer, while the rest of the network that communicates with the memory banks is divided in $Nx\frac{M}{K}$ smaller networks distributed on each memory layer. We first explore the impact of the 3D partitioning on the network area, measured as equivalent kgates (nand2), for several systems:

- 16 PEs and 64MMs.
- 16 PEs and 128MMs.
- 8 PEs and 64MMs.
- 8 PEs and 128MMs.

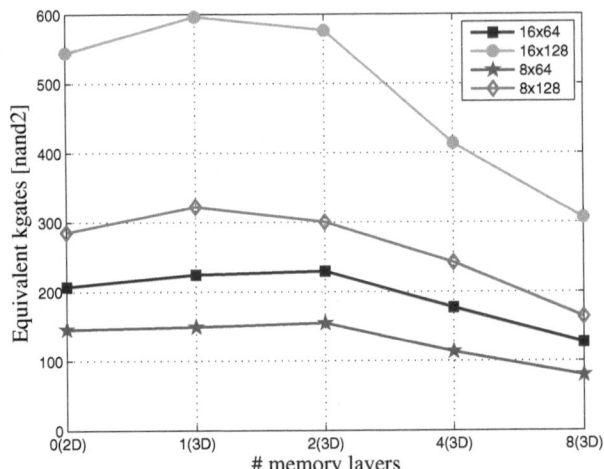

Fig. 4. Area occupied by the network in the 3D system

Figure 4 depicts the trend of the total area, that is the sum of the area occupied by the partitioned network on each layer, for different network cardinalities. We can notice that for 3D-systems composed of 1 memory layer, the total area has a slight increase. This is due to the fact that moving from a 2D-system to a 3D-system, the small stall network is added on the logic layer. Once we reach 3 or more layers, even if the network is replicated on each memory layer, the area reduction per layer dominates. Since the total number of primitives constituting 3D-LIN is equal to $\sum_{j=1}^{K} \sum_{i=1}^{log_2 \frac{M}{K}} \frac{\frac{M}{K}}{2^i} \times N + \sum_{j=1}^{K} \sum_{i=1}^{log_2 N} \frac{M}{K} \times \frac{N}{2^i}$, is expected that the area reduction is more accentuated for networks connecting a higher number of MMs.

In a 3D system, however, is important to consider the per-layer reduction, since the form factor is influenced by the single layers dimension. The area occupied by the network on the logic layer and the ones on each memory layer is shown in Figure 4.1. Once adding more memory layers, there is a strong decrease in the per-layer network area.

Figure 6 shows the trend of the ratio between the network area and the memory area both per layer and in the full 3D system composed of 16 PEs interfaced to 64 MMs. When moving from a planar design to a stacked system, the sum of the ne twork areas on each layer is higher than the 2D counterpart, nevertheless the area per layer decreases.

The configurability of the Network gives the possibility to explore the form-factor trend for the 3D multi-core systems with shared L1 memory on top of logic. Given the specification of the system, the best trade-off can be found in terms of number of layers. In particular, we chose to analyze the area of the chip(A_{3D}) normalized to the area of the same chip implemented on a single

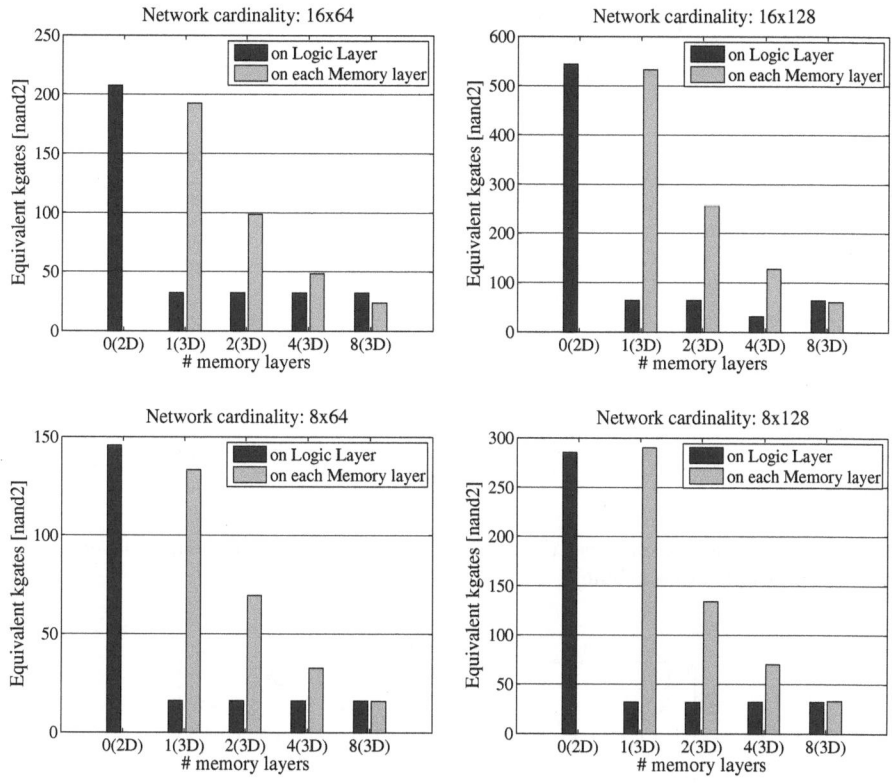

Fig. 5. Area of the Stall/Valid Network on the logic layer (blue) and area of the data Network on each memory layer (green) for different number memory layers stacked on top of the logic layer

silicon layer(A_{2D}) for the following configurations and area occupation of the memory(A_{mem}) over the area of the planar chip(A_{2Dchip}):

- 16 PEs and 16 MMs : $\frac{A_{mem}}{A_{2Dchip}}$=43% ;
- 16 PEs and 32 MMs : $\frac{A_{mem}}{A_{2Dchip}}$=58%;
- 16 PEs and 64 MMs : $\frac{A_{mem}}{A_{2Dchip}}$=70% ;
- 16 PEs and 128 MMs : $\frac{A_{mem}}{A_{2Dchip}}$=79% .

Figure 7 depicts the reduction of the area when the chip is designed to stack different numbers of memory layers on top of the logic layer. When moving from the planar structure, to a 2-layer structure, the memories and the network are moved to the upper layer, and we can notice a decrease in the form factor. However, this reduction is still limited due to the size of the network that, as explained before, does not shrink effectively. In additions, the TSV area occupation increases the size of both layers. Considering the stacking of two or more layers on top of the logic, the network cardinality start changing depending on

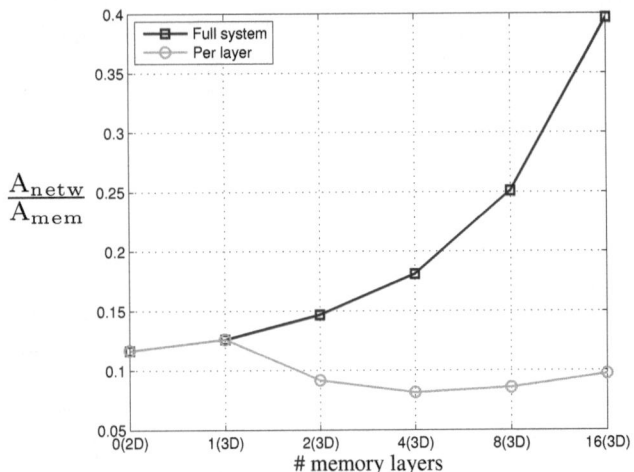

Fig. 6. Area of the network over the area of the memory for each memory layer(green), and for the whole system(blue)

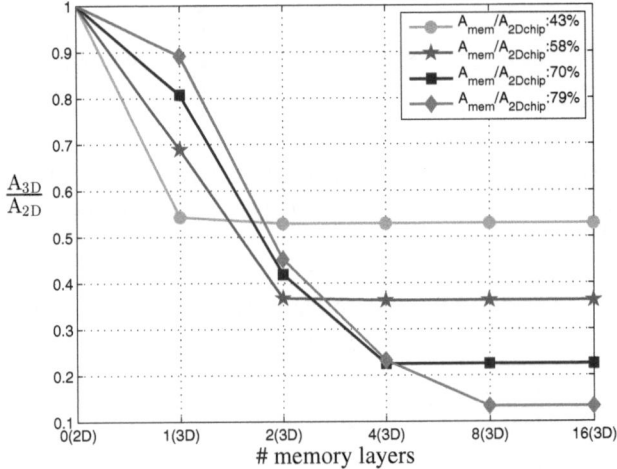

Fig. 7. Area of the 3D chip normalized to the area of the 2D implementation

the number of memory layers, leading to a decrease in its area occupation, while the TSV occupation remains the same as for the 3D, single memory layer, case. The best trade-off point can be found when the area of the memory layer is almost equal to the area of the logic layer. When reaching the best trade-off, the stacking of any more memory layers does not affect the form factor that is now defined from the area of the logic layer.

4.2 Power Analysis

The power consumption is an important parameter to be considered. For 3D-ICs, it is even more important: stacking more layers arise new challenges due to an increased power density per footprint, which may cause temperature to increase beyond the limits that guarantees reliability. At the design level, careful floorplan definition and thermal management techniques such as *dynamic voltage and frequency scaling (DVFS)* can help, but are not sufficient. There is a significant research effort to tackle the power issue at different levels. At the software level thermal-aware task scheduling policies [23] can be implemented, while at the fabrication level, cooling techniques such as inter-layer micro-channel liquid cooling [22] and *Thermal-TSVs(TTSV)* [24], [25] can be exploited to remove the excessive heat.

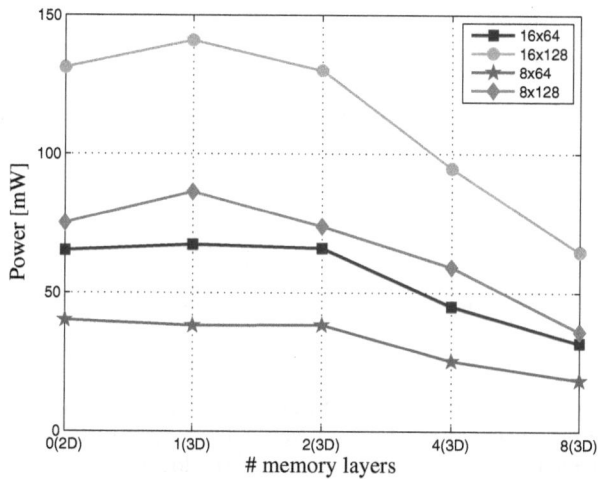

Fig. 8. Total dynamic power consumption of the network in the 3D system

In this chapter, we do not propose any cooling or thermal management techniques, but we focus on exploring the power dissipation of 3D-LIN to ensure reliability. The total dynamic power consumed by the network is depicted in figure 8. We can observe how the trend for power is correlated to the network area. As the number of blocks to be interconnected increases, the size of the die affect the wire length and the power related to wiring start dominating the cell internal power. Hence, the gain in power consumption is more pronounced for systems with higher cardinality and appears once stacking more memory layers which reduces both the per-layer network cardinality, and the single layer size.

The power contribution of the different single layers is shown in figure 8. The power consumed by the stall network on the logic layer is small compared to the consumption of the network on each memory layer, which is the dominant contribution. As the number of stacked memory layers increases, the cardinality of the network on each layer is reduced, leading to a significant gain in power.

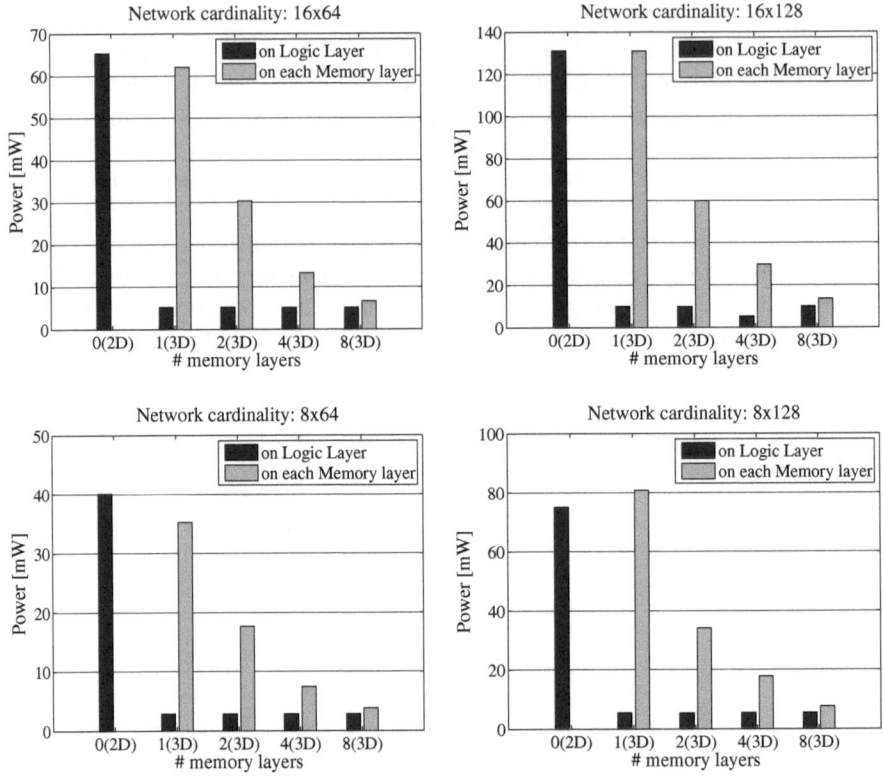

Fig. 9. Dynamic power consumed by the Stall/Valid Network on the logic layer (blue) and dynamic power consumed by the data Network on each memory layer (green) for different number memory layers stacked on top of the logic layer

4.3 Timing Analysis

Exploring 3D-LIN in term of latency the following configurations are considered:

- 16 PEs and 32 MMs;
- 16 PEs and 64 MMs;
- 16 PEs and 128 MMs.

As previously discussed, the frequency of the network is limited by the response path that includes the access time to read a data from the memory bank. However, depending on the size of the memory module, this access time changes. In our experiments, we explored the latency of the network when connecting memory banks of 8kB. In Figure 4.3 and 4.3, both system latency and network latency are shown. We can notice that moving from the planar system to one stacked memory layer, the latency slightly decreases due to the shorter interconnect. The reduction in delay is more evident for the systems with two

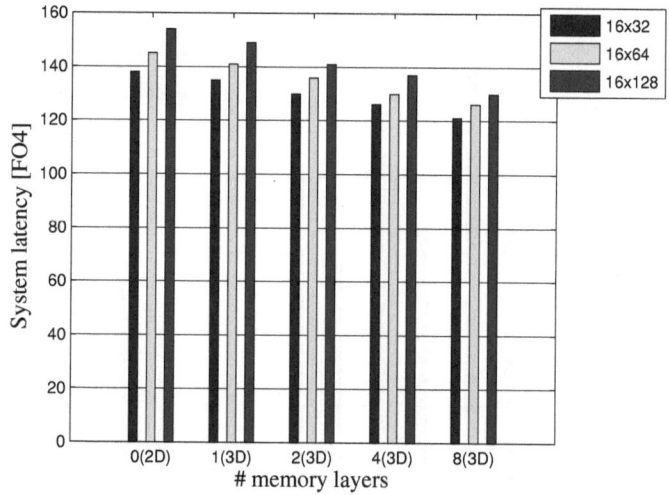

Fig. 10. System latency: Network delay plus memory access time

or more memory layers, due to the changes in the network topology. The reduction in delay is more evident in Figure 4.3 considering the network itself, independently from the attached memory banks. The latency of the network shows significant improvement, in the case of 16PEs connected to 64MMs, the 2D latency of ˜42FO4 is reduce down to ˜23FO4 .

Table 2 shows the latency improvements in percentage. The results show that stacking a single memory layer, the memory access time dominates the decreased latency of the interconnect and the improvement is only a few percents. However, when we move to two memory layers, we can obtain already around 8% improvement, reaching 11% with four memory layers for a network cardinality of 16x128. Independently from the attached memory, considering the network alone, the benefits are more evident, with 35% improvements for four memory layers stacked on top of the logic layer.

Table 2. Latency improvement

	16x32		16x64		16x128	
	system	network	system	network	system	network
1 memory layer	2%	9%	2%	7%	3%	10%
2 memory layers	6%	22%	6%	20%	8%	24%
4 memory layers	8%	32%	10%	35%	11%	31%
8 memory layers	12%	46%	13%	44%	16%	46%

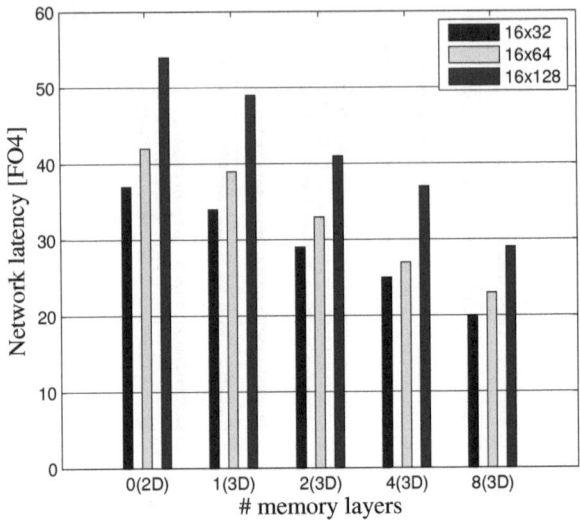

Fig. 11. Network latency

5   Conclusion

In this paper, we present a configurable network architecture that can be integrated in 3D stacked CMP. The network enable the connection of multiple processing elements to a shared multi-banked memory guaranteeing low-latency connection. The network and the multi processor system has been explored in terms of area, form factor, power and latency. The benefits obtained by exploiting 3D integration are evaluated. Moreover, the study also focus on exploring the performances for different 3D structures, studying the effects of stacking different number of layers. The physical synthesis results show the best trade off point between the amount of memory needed in the system and the number of stacked layers. In case of a memory occupation of 60% of the planar chip, by moving to a system that integrates two memory layers on top of a logic layer, the form factor is improved more than 60%. In terms of latency, the 16x128 configuration of the network can be improved up to around 24% in case of 2 memory layers, and 31% in case of four memory layers, leading to a latency reduction for accessing 8kB memory banks of 8% and 11% respectively. Latency and area improvements come without a worsening in terms of power. Stacking 2 or 3 layers, the power consumption is kept almost the same as for the 2D implementation, while starts improving as the number of layer increases.

Acknowledgments. This work has been partially supported by the EU project grant PRO3D FP7-ICT-248776.

References

1. Owens, J.D., Dally, W.J., Ho, R., Jayasimha, D.N., Keckler, S.W., Peh, L.-S.: Research challenges for on-chip interconnection networks. IEEE Micro 27, 96–108 (2007)
2. Borkar, S., Chien, A.A.: The Future of Microprocessors. Commun. ACM 54, 67–77 (2011)
3. Benini, L., De Micheli, G.: Networks on Chips: a New SoC Paradigm. Computer 35, 70–78 (2002)
4. Balkan, A., Qu, G., Vishkin, U.: A Mesh-of-Trees Interconnection Network for Single-Chip Parallel Processing Application-Specific Systems. In: International Conference on Architectures and Processors, pp. 73–80 (2006)
5. Plurality, Ltd.: The hyperCore architecture. White Paper (2010)
6. Rahimi, A., Loi, I., Kakoee, M., Benini, L.: A fully-synthesizable single-cycle interconnection network for Shared-L1 processor clusters Design. In: Automation Test in Europe Conference, pp. 1–6 (2011)
7. Xie, Y.: Processor Architecture Design Using 3D Integration Technology. In: 23rd International Conference on VLSI Design, pp. 446–451 (2010)
8. Li, F., Nicopoulos, C., Richardson, T., Xie, Y., Narayanan, V., Kandemir, M.: Design and management of 3D chip multiprocessors using network-in-memory. SIGARCH Comput. Archit. News 34, 130–141 (2006)
9. Loh, G.: 3D-Stacked memory architectures for multi-core processors. In: Proceedings of the 35th Annual International Symposium on Computer Architecture, pp. 453–464 (2008)
10. Woo, D.H., Seong, N.H., Lewis, D., Lee, H.-H.: An Optimized 3D-Stacked Memory Architecture by Exploiting Excessive, High-Density TSV Bandwidth. In: 16th International Symposium on High Performance Computer Architecture, pp. 1–12 (2010)
11. Madan, N., Zhao, L., Muralimanohar, N., Udipi, A., Balasubramonian, R., Iyer, R., Makineni, S., Newell, D.: Optimizing communication and capacity in a 3D stacked reconfigurable cache hierarchy. In: 15th International Symposium on High Performance Computer Architecture, pp. 262–274 (2009)
12. Mishra, A., Dong, X., Sun, G., Xie, Y., Vijaykrishnan, N., Das, C.: Architecting on-chip interconnects for stacked 3D STT-RAM caches in CMPs. SIGARCH Comput. Archit. News 39, 69–80 (2011)
13. Li, F., Nicopoulos, C., Richardson, T., Xie, Y., Narayanan, V., Kandemir, M.: Design and Management of 3D Chip Multiprocessors Using Network-in-Memory. SIGARCH Comput. Archit. News 34, 130–141 (2006)
14. Kim, J., Nicopoulos, C., Park, D., Das, R., Xie, Y., Narayanan, V., Yousif, M., Das, C.: A novel dimensionally-decomposed router for on-chip communication in 3D architectures. In: 34th International Symposium on Computer Architecture, pp. 138–149 (2007)
15. Park, D., Eachempati, S., Das, R., Mishra, A., Xie, Y., Vijaykrishnan, N., Das, C.: MIRA: A Multi-layered On-Chip Interconnect Router Architecture. In: 35th Annual International Symposium on Computer Architecture, pp. 251–261 (2008)
16. Xu, Y., Du, Y., Zhao, B., Zhou, X., Zhang, Y., Yang, J.: A Low-Radix and Low-Diameter 3D Interconnection Network Design. In: 15th International Symposium on High Performance Computer Architecture, pp. 30–42 (2009)
17. Xue, L., Gao, Y., Fu, J.: A High Performance 3D Interconnection Network for Many-Core Processors. In: 2nd International Conference on Computer Engineering and Technology, pp. 383–389 (2010)

18. Ben Ahmed, A., Ben Abdallah, A., Kuroda, K.: Architecture and Design of Efficient 3D Network-on-Chip (3D NoC) for Custom Multicore SoC. In: Broadband, Wireless Computing, Communication and Applications, pp. 67–73 (2010)
19. Design Compiler User Guide, Synopsys, version F-2011.09-SP2 (2011)
20. Van der Plas, G., Limaye, P., Loi, I., Mercha, A., Oprins, H., Torregiani, C., Thijs, S., Linten, D., Stucchi, M., Katti, G., Velenis, D., Cherman, V., Vandevelde, B., Simons, V., De Wolf, I., Labie, R., Perry, D., Bronckers, S., Minas, N., Cupac, M., Ruythooren, W., Van Olmen, J., Phommahaxay, A., de Potter de ten Broeck, M., Opdebeeck, A., Rakowski, M., De Wachter, B., Dehan, M., Nelis, M., Agarwal, R., Pullini, A., Angiolini, F., Benini, L., Dehaene, W., Travaly, Y., Beyne, E., Marchal, P.: Design issues and considerations for low-cost 3-D TSV IC technology. J. of Solid-State Circuits 46, 293–307 (2011)
21. Kim, D.H., Mukhopadhyay, S., Lim, S.K.: Fast and Accurate Analytical Modeling of Through-Silicon-Via Capacitive Coupling. IEEE Transactions on Components Packaging and Manufacturing Technology 1, 168–180 (2011)
22. Shi, B., Srivastava, A.: Liquid Cooling for 3D-ICs. In: International Green Computing Conference and Workshops, July 25-28, pp. 1–6, (2011)
23. Zhou, X., Yang, J., Xu, Y., Zhang, Y., Zhao, J.: Thermal-aware Task Scheduling for 3D Multicore Processors. IEEE Trans. Parallel Distrib. Syst. 21, 60–71 (2010)
24. Goplen, B., Sapatnekar, S.: Thermal Via Placement in 3D ICs. In: International Symposium on Physical Design, pp. 167–174 (2005)
25. Yu, H., He, L.: Dynamic Power and Thermal Integrity in 3D Integration. In: Communications, Circuits and Systems, pp. 1108–1112 (2009)

A Hexagonal Processor and Interconnect Topology for Many-Core Architecture with Dense On-Chip Networks

Zhibin Xiao and Bevan Baas

Department of Electrical and Computer Engineering
University of California, Davis
1 Shields Avenue Davis, CA USA 95616
{zxiao,bbaas}@ucdavis.edu

Abstract. Network-on-Chips (NoCs) are used to connect large numbers of processors in many-core processor architecture because they perform better than less scalable methods such as global shared buses. Among all NoC design parameters, NoC topologies define how nodes are placed and connected and greatly affect the performance, energy efficiency, and circuit area of many-core processor arrays. Due to its simplicity and the fact that processor tiles are traditionally square or rectangular, 2D mesh is mostly used for existing on-chip networks. However, efficiently mapping applications can be a challenge for cases that require communication between processors that are not adjacent on the 2D mesh. Motivated by the fact that applications often have largely localized communication patterns, we have proposed an 8-neighbor mesh topology and a 6-neighbor topology with hexagonal-shaped processor tiles, both of which increase local connectivity while keep much of the simplicity of a mesh-based topology. We have physically designed a 16-bit DSP processor and the corresponding processor arrays which utilize all three topologies. A 1080p H.264/AVC residual video encoder and a 54 Mbps 802.11a/11g OFDM wireless LAN baseband receiver are mapped onto all topologies. The 6-neighbor hexagonal grid topology incurs a 2.9% area increase per tile compared to the 4-neighbor 2D mesh, but its much more effective inter-processor interconnect yields an average total application area reduction of 21%, an average power reduction of 17%, and a total application inter-processor communication distance reduction of 19%.

Keywords: CMOS, many-core processor, interconnection topology, network on chip (NoC), digital signal processing (DSP).

1 Introduction

Tiled architectures that integrate two or more independent processor cores are called multi-core processors. Manufactures typically integrate multi-core processors into a single integrated circuit die (known as chip multiprocessors or CMP). CMPs that integrate tens, hundreds, or thousands of cores per die are called

A. Burg et al. (Eds.): VLSI-SoC 2012, IFIP AICT 418, pp. 125–143, 2013.

many-core chips and those that utilize scalable interconnects and avoid long global wires will attain higher performance [1].

NoCs are used to connect large numbers of processors in many-core processor architecture because they perform better than less scalable methods such as global shared buses. Among all NoC design parameters, NoC topologies define how nodes are placed and connected and greatly affect the performance, energy efficiency, and circuit area of many-core processor arrays. Due to its simplicity and the fact that processor tiles are traditionally square or rectangular, 2D mesh is mostly used for existing on-chip networks. However, efficiently mapping applications can be a challenge for cases that require communication between processors that are not adjacent on the 2D mesh as shown in Figure 1(a). This condition could require processors to act as routing processors for static interconnection architectures, and intermediate routers for dynamic router-based NoCs. The power consumption and communication latency also increase as the number of routing processors or routers between two communicating cores increase.

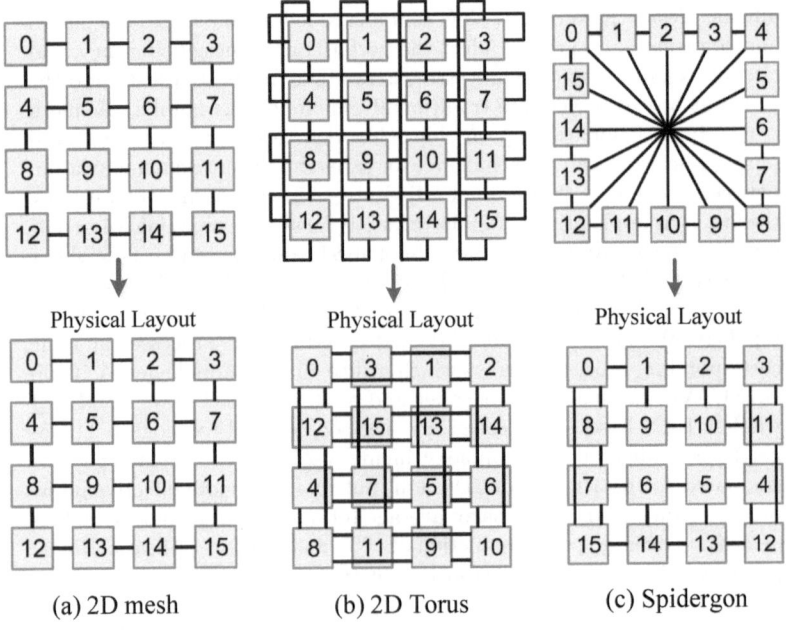

Fig. 1. Popular Network-on-Chip topologies and their physical layouts [2]: (a) 2D mesh, (b) 2D Torus, and (c) Spidergon

There exist other common topologies for NoCs such as 2D torus, Spidergon, fat tree and higher dimensional meshes and tori which provide higher routing capability and communication bandwidth with costs of higher wire density and longer global wires. Furthermore, topologies with irregular layouts present significant challenges for many-core implementations especially with the number of cores per die expected to soon reach thousands and more. As an example, Figure 1(b)(c) shows the 2D torus and Spidergon topologies as well as their physical layouts on a 2-dimensional chip [2]. Both topologies require global wires which go across one or more processors. Mapping arbitrary non-regular topologies to a 2D floorplan is an NP-hard optimization problem [3].

For many applications mapped onto homogeneous chip multiprocessors, communication within processors is often largely localized [4], which may result in local mapping congestion. An increase of local connectivity can ease such congestion, which results in application mappings with smaller application area and lower power consumption. This motivates us to propose new topologies with increased local connectivity while keeping much of the simplicity of a mesh-based topology.

The main contributions of this paper can be summarized as three points. First, we have proposed a 6-neighbor topology with hexagonal-shaped processor tiles and a 8-neighbor mesh topology, which are compared to the common 4-neighbor 2D mesh topology. Second, commonly available commercial CAD tools are used to implement tiled CMPs for all three topologies. Three processors including a hexagonal-shaped processor tile and their corresponding many-core processor arrays are physically implemented in 65 nm CMOS and are DRC and LVS clean. Third, a complete functional H.264/AVC residual encoder and an 802.11a baseband receiver are mapped onto all three topologies for realistic comparisons.

The remainder of this paper is organized as follows. Section 2 describes the related work. Section 3 presents the proposed inter-processor communication topologies. Section 4 shows the mapping of two complex applications to all discussed topologies. In section 5, the physical design of the hexagonal-shaped processor tiles is presented. Section 6 presents the chip implementation results and section 7 concludes this paper.

2   Related Work

Many topologies have been used for on-chip inter-processor communication, such as buses, meshes, tori, binary trees, octagons, hierarchical buses and custom topologies for specific applications. The low complexity 2D mesh has been used by most fabricated many-core systems including RAW [5], AsAP [6], TILE64 [7], AsAP2 [8] and Intel 48-core Single-Chip Cloud Computer (SCC) [9].

Becker et al. [11] developed a hexagonal Field-programmable Analog Array in a 0.13 μm CMOS technology. The basic building block is a hexagonal analog circuit block which communicates with six neighbors. Extension to a many-core processor is similar in topology, but very different in terms of impact on tile area and total application interconnect.

Malony studies the two-dimensional regular processor arrays which are geometrically defined based on nearest-neighbor connections and space-filling properties [12]. He theoretically proves the hexagonal array is the most efficient topology in emulating other topologies by analyzing the geometric characteristics. Chen et al. theoretically explored the addressing, routing and broadcasting

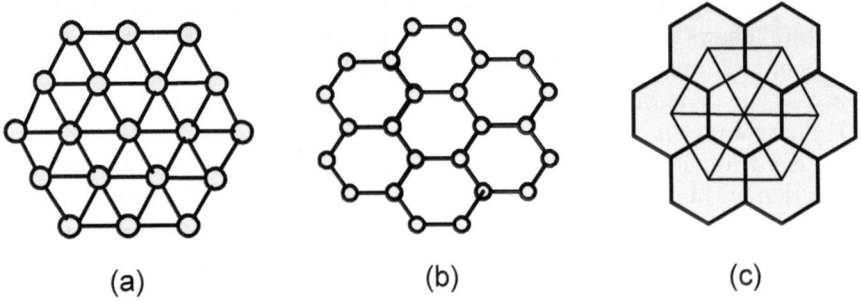

Fig. 2. Examples of three hexagonal networks (a) a 6-neighbor off-chip hexagonal network; (b) a 3-neighbor on-chip honeycomb network [10]; (c) the proposed 6-neighbor on-chip hexagonal grid network

in hexagonal mesh multiprocessors [13]. Decayeux and Seme proposed a 3D hexagonal network as an extension of 2D hexagonal networks [14]. Their work focuses on off-chip 6-neighbor hexagonal network where each node is located at the vertex of the network as shown in Figure 2(a). Stojmenovic proposed efficient coordinate system and routing algorithms for the 3-neighbor honeycomb mesh networks as shown in Figure 2(b) [15]. Compared to previous work, we have designed a hexagonal-shaped processor that can be tiled together as a hexagonal mesh for on-chip inter-processor communication as shown in Figure 2(c). The advantages of hexagonal-shaped processor topology are demonstrated by real-world application mappings and physical implementations of a fully functional many-core processor array.

3   Processor Shapes and Topologies

3.1   NoC Topology Analysis Criteria

NoC topologies can be analyzed by a few criteria [16]:

- *Degree*: is the number of direct neighbors for one node. A high degree allows more nodes to communicate directly with low latency.
- *Diameter*: is the largest number of hops between any two nodes. A small diameter indicates low maximum latency of a network.
- *Bisection*: is the minimum number of links to be removed to separate a network into two equal ones. A high bisection indicates a high bandwidth yielding high throughput.
- *Number of links*: the total number of bidirectional links in a network.
- *Clustering degree*: also called clustering coefficient, is a measure of degree to which nodes in a network tend to cluster together. The local clustering degree for a node i can be defined as: $\frac{2l_i}{n_i(n_i-1)}$, where n_i is the number of direct neighbors and l_i is the number of links between its neighboring nodes. A high clustering degree indicates that local nodes close to each other are strongly connected.

Table 1. Characteristics of various regular topologies for a homogenous many-core array with $n \times n$ processors where n is the number of processors on one edge and $n \geq 2$

Topology	Degree	Max. Link Hops	Link Num.	Diameter	Bisection	Clustering Degree
2D Mesh	4	0	$2n(n-1)$	$2(n-1)$	n	0
2D Torus	4	1	$2n^2$	n	$2n$	0
8-neighbor mesh	8	1	$4n^2 - 6n + 2$	$n - 2$	$3n - 2$	0.86
6-neighbor hexagon	6	0	$3n^2 - 4n + 1$	$n + \lfloor \frac{n-2}{2} \rfloor$	$2n - 1$	0.40

$^+$ Omitted due to space limitation. The total number of links for 5-5 House and Rect is: $n(n-1) + n(\lfloor \frac{n-1}{2} \rfloor) + (2n-1)(\lceil \frac{n-1}{2} \rceil)$.

* This is for $n \geq 4$. If $n \leq 3$, the diameter of the topology is: $n + \lfloor \frac{n-1}{2} \rfloor$.

– *Max link hops*: is the maximum hops that a link can cross after the topology has been physically mapped to a 2-dimensional chip. This is a criteria proposed in this work to measure the length of global wires of a topology.

The above criteria can be used to compare various topologies and provide an initial indication on performance. The first two rows of Table 1 list the characteristics of two popular topologies 2D mesh and 2D torus. 2D mesh has a maximum degree of 4 and a maximum link hop equal to 0 since all of the links are nearest-neighbor. For an $n \times n$ array, 2D mesh has a number of links equal to $2n(n-1)$, a diameter equal to $2(n-1)$, bisection n , and a clustering degree equal to 0. Compared with 2D mesh, 2D torus has the same degree, more links, smaller diameter, higher bisection bandwidth and the same clustering degree. All of these criteria indicate 2D torus could achieve higher throughput and lower latency at the cost of more long non-nearest neighbor links.

This work explores low complexity topologies with higher degree, larger number of links, smaller diameter, higher bisection compared to 2D mesh. We also limit the maximum link hops being less than or equal to one to avoid global long wires. These requirements result in proposed topologies that have a strong local connectivity with a non-zero clustering degree. In the following subsections, a 6-neighbor hexagon and an 8-neighbor mesh topologies are proposed and analyzed.

3.2 Processor Tile Shapes

To the best of our knowledge, all previously-fabricated VLSI processors have been of a rectangular shape, often nearly square. As illustrated in Figure 3(a)(b), it stands to reason that a circular shape would allow shorter wires for a given netlist, resulting in smaller area and lower wire capacitance which would result

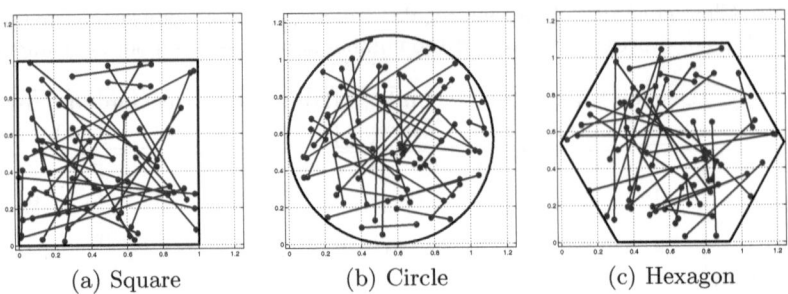

(a) Square   (b) Circle   (c) Hexagon

Fig. 3. Example tiles of constant area with random uniformly-distributed wire endpoints

in higher speeds and lower energy per operation. A simple experiment with ideal shapes and one million randomly-placed wires yields a 2.2% reduction in total wire length for a circular tile compared to a square tile. On the negative side, it is clear that circles do not pack together without wasted space between tiles. On the positive side, circles pack with *six* neighbors while rectangles obviously have only four. It is reasonable to expect a rectangular tile to have longer wires on average compared to a square tile.

In contrast to the circle, the hexagonal shape *does* pack efficiently without gaps between tiles and it retains the 6-nearest-neighbor property. The same wiring experiment was run for a hexagonal tile and it resulted in a 1.8% reduction in total wire length compared to the square tile. A reduction in total wire length yields a pure benefit in area, energy and delay for processor tile design. The inclusion of common rectangular blocks such as memory arrays in a processor tile increases routing congestion but is shown in Section 6 to be tolerable. In addition, we demonstrate that Manhattan-style wire routing is fully compatible with non-rectangular tile shapes.

3.3 The Proposed Topologies

As shown in Fig. 4, three different topologies are studied and the well-known 4-neighbor mesh is used as the baseline topology for comparison as shown in Figure 4(a).

Figure 4(b) shows a 6-neighbor processor array using hexagonal-shaped processor tiles. The processor center-to-center distance is $\sqrt{3} * w$ if the length of the hexagon edge is w. The hexagonal grid is commonly used in mobile wireless networks due to its desirable feature of approximating circular antenna radiation patterns and its optimal characteristic of six nearest neighbors. The symmetry and space-filling property make the hexagonal-shaped processor tile an attractive design option for many-core processor arrays.

Due to limitations of current wafer sawing machines, chips on round wafers are traditionally square or rectangular. In fact, the opportunities and limitations of non-rectangular processors on a chip are analogous to non-rectangular chips

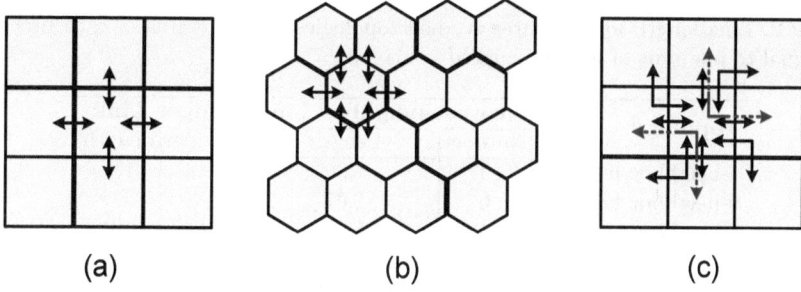

Fig. 4. The three inter-processor communication topologies considered in this work: (a) baseline 4-neighbor mesh (b) 6-neighbor hexagonal tile and interconnect, and (c) 8-neighbor mesh

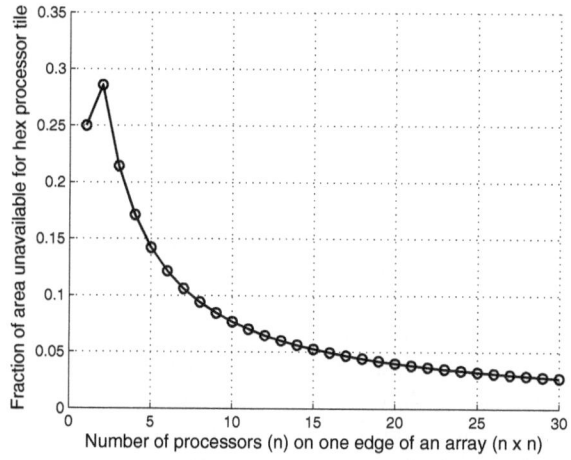

Fig. 5. Fraction of area unavailable for processor tiles in an n x n many-core array utilizing 6-neighbor hexagonal tiles and interconnect topology

on a wafer. For the case of a rectangular chip composed of hexagonal-shaped processors, there are areas on the periphery of the chip in which processors can not be placed. Figure 5 shows the percentage of unavailable area for the hexagonal-shaped tile topology with varying processor array size. If the processor array size is larger than 30 x 30, this area overhead becomes less than 2.7% of the total chip area. In practice, this area could be filled with other chip components such as decoupling capacitors, or portions of hardware accelerators, memory modules, I/O circuits or power conversion circuits.

Another logical extension of the 2D mesh is to include four diagonal processors in an 8-neighbor arrangement as shown in Figure 4(c) where each rect tile can directly communicate with 8 neighbors. This approach has increased routing congestion in the tile corners due to the four (uni-directional) links that pass through each corner (the dashed lines in Figure 4(c)).

Table 2. Link length for the three studied topologies with the area of each processor tile equal to one unit of length squared

Topology	Nearest-neighbor Link		Longer Link	
	Number	Length	Number	Length
4-neighbor mesh	4	1.00	–	–
6-neighbor hex grid	6	1.07	–	–
8-neighbor mesh	4	1.00	4	1.41

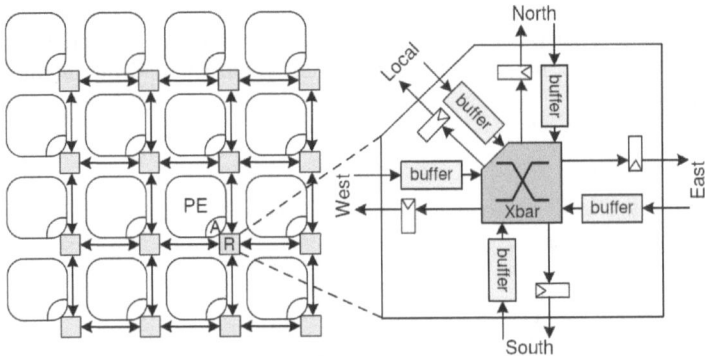

Fig. 6. A 2D mesh processor array connected by a dynamic five-port routers each with one port connected to the processor core

Table 1 also lists the characteristics of the two proposed topologies for an homogenous many-core array with $n \times n$ processors. Compared with 2D mesh, the two proposed topologies have larger node degree, smaller diameter, larger or equal bisection bandwidth and larger clustering degree. The 8-neighbor mesh topology has the largest bisection, smallest diameter and largest clustering degree, which indicates lower maximum latency and high maximum throughput. The advantage of the 6-neighbor hexagon topologies is that global long wires are not required.

The center-to-center distance can be used to represent the communication link length between two "touched" processors. Table 2 shows the number of different types of communication links and the corresponding link length for the three topologies. The area of a processor tile is assumed to be one squared unit. As shown in Table 2, the 4-neighbor mesh and 6-neighbor hex grid have only one type of communication link due to equal center-to-center tile distance. The 8-neighbor mesh topology has two types of links.

4   Application Mapping

4.1   Target Interconnect Architecture

Fig. 6 shows the inter-processor communication in a typical 2D mesh processor array using dynamic routers. As the diagram shows, the processor array is

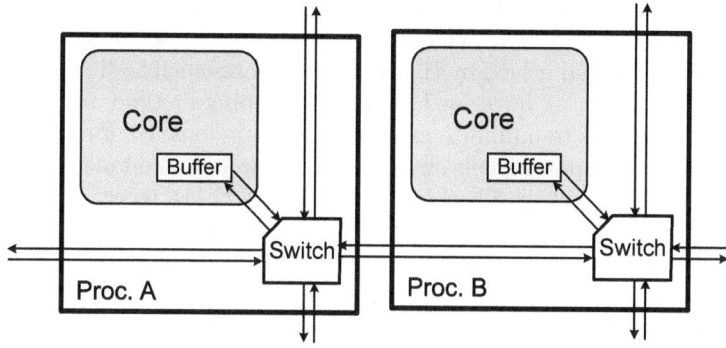

Fig. 7. A 2D mesh processor array connected by circuit switches each with four nearest-neighbor inter-processor communication links and one port connected to the processor core

connected by 5-port routers and the communication logic includes five buffers and one 5 × 5 crossbar. There might be more control logic to support the communication flow control which is not drawn. The static circuit-switch interconnection has smaller area, lower power dissipation and lower complexity than dynamic router interconnection while trading off routing flexibility [17].

Fig. 7 shows another 2D mesh array connected by circuit switches each with four nearest-neighbor interconnection links and one port connected to the processor core. The circuit switch communication logic has only one buffer and one 4 × 1 crossbar. The long distance communication is performed by software in the intermediate processors. In this work, we use the static configurable circuit switch architecture which is suitable for applications with steady communication patterns. We also extend the architecture in Fig. 7 by adding one more port to the processor core due to the fact that processors normally have a two-operand instruction format. Thus, the processor can read two words from two buffers in one instruction at the same time.

4.2   Application Mapping Methodology

Parallel programming on the discussed many-core systems with dense on-chip networks includes two main steps: 1) partitioning the algorithms at a fine-grained level; 2) mapping the tasks to the nodes of the processor array and connecting the nodes with available links defined by the topology [18]. The two steps might be repeated iteratively for throughput optimization where we can identify the bottleneck task of the design and partition it even more until the throughput meets the requirement.

To be specific, in the partitioning step, an estimate of task workload and required resources such as data and instruction memories are used to generate a fine-grained task graph where each task can be assigned to one processor node. Following the fine-grained partition, the mapping is conducted either manually or automatically by an automatic mapping tool [19]. Application mapping is essentially an optimization problem, which can be formed as integer linear programming (ILP) problem [20] and solved by Heuristic algorithms such as simulated annealing [21]. In this work, we have used a manual mapping method and the primary optimization target is to minimize area and maximize local communication.

Based on the two-port circuit switch architecture, two complete applications including an H.264/AVC residual encoder and an 802.11a receiver are manually mapped onto all three topologies which differ in the number of links among neighboring processor tiles. In order to be fair to compare all topologies, we chose not to partition tasks specifically for one topology and mapping the two applications onto all topologies is based on the same task graph.

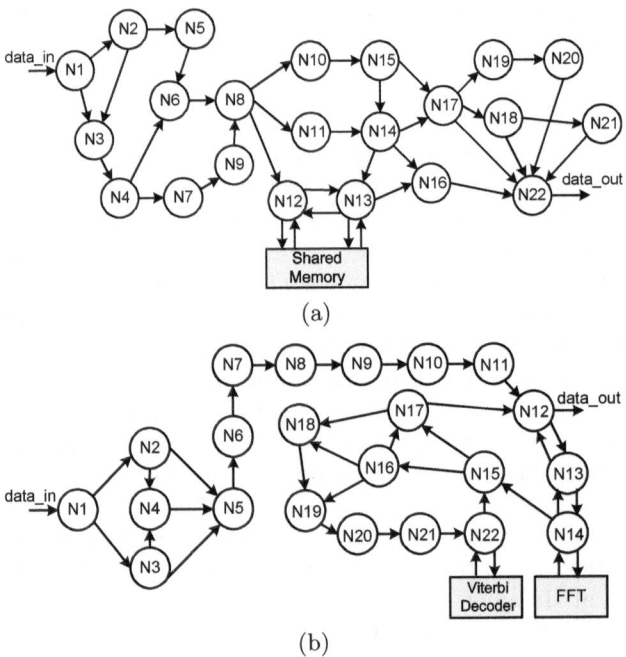

Fig. 8. Task graph of (a) a 22-node H.264/AVC video residual encoder, and (b) a 22-node 802.11a WLAN baseband receiver

4.3 Benchmark Application Mapping

Figure 8 depicts two task graphs of the benchmark applications, where each node represents one task which can be implemented in one processor and each edge represents one physical link between two processor nodes. Figure 8(a) shows a 22-node task graph of an H.264/AVC residual baseline encoder composed of integer transform, quantization and context-adaptive and variable length coding (CAVLC) functions [18]. The H.264/AVC encoder is a memory-intensive application which requires an additional shared memory module as shown in the task graph. Figure 8(b) shows a 22-node task graph of a complete 802.11a WLAN baseband receiver which is computation-intensive requiring two dedicated hardware accelerators: Viterbi decoder and FFT. The complete receiver includes necessary practical features such as frame detection, timing synchronization, carrier frequency offset (CFO) estimation and compensation, and channel estimation and equalization [22]. Figure 9 shows an example mapping of the

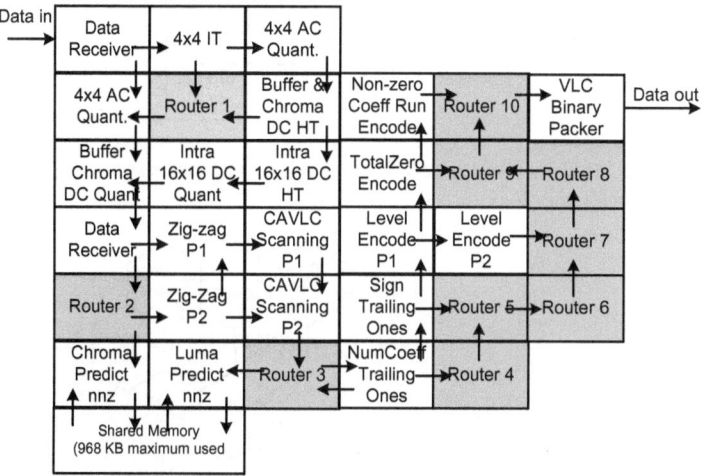

Fig. 9. An H.264/AVC video residual encoder mapped on a processor array with 4-neighbor 2D mesh topology

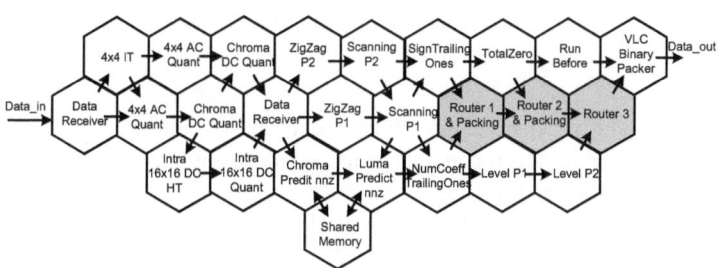

Fig. 10. An H.264/AVC residual video encoder mapped on a processor array with 6-neighbor hex topology

H.264/AVC residual encoder capable of 1080p HDTV encoding at 30 frames per second on the baseline 4-neighbor mesh that uses 32 processors plus one shared memory. The 4-neighbor mesh is inefficient in handling a complex application like H.264/AVC encoding. A total of 10 processors are used solely for routing data which accounts for 31% of the total application area. Figure 10 shows a possible 25-processor mapping on the proposed 6-neighbor hex grid topology. As mentioned before, the hexagonal-shaped processors still take a maximum of

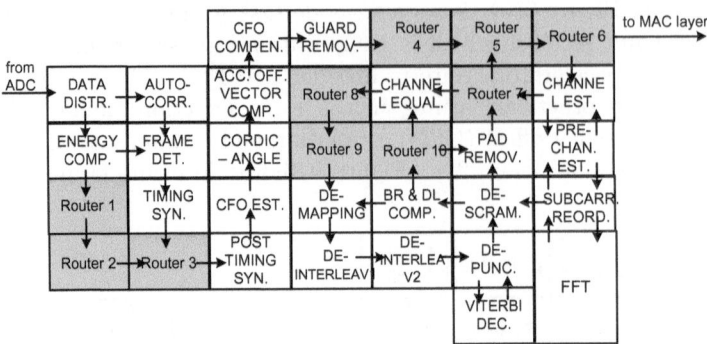

Fig. 11. An 802.11a baseband receiver mapped on the processor array with baseline 4-neighbor 2D mesh topology

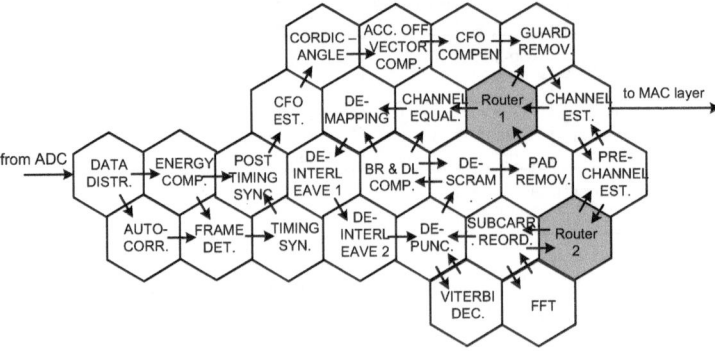

Fig. 12. An 802.11a baseband receiver mapped on the processor array with 6-neighbor hex topology

two inputs from the six nearest-neighbor processors. Compared with the design using 4-neighbor mesh, seven routing processors are saved, which accounts for a 22% processor number reduction.

Fig. 11 shows a mapping of the 802.11a/g baseband receiver (54 Mbps) on the baseline 4-neighbor 2D mesh that uses 32 processors plus the Viterbi decoder and FFT accelerators with 10 processors used for merging and forwarding data. Fig. 12 shows a mapping on the hexagonal-shaped tile architecture which requires only 24 processors plus the Viterbi decoder and FFT accelerators—25% fewer processors than those used in the 2D mesh mapping.

4.4   Application Mapping Results

Figure 13(a) shows the number of processors used for mapping the two applications to all three topologies. The 6-neighbor hex grid and 8-neighbor mesh are much more efficient than the baseline 2D mesh, resulting in a number of processor savings of 25% and 22% for the H.264 residual encoder and both 25%

for the 802.11a receiver. The 8-neighbor mesh requires slightly larger number of processors than the 6-neighbor hex grid topology which yields the largest reduction (24%) in average number of used processors compared to 4-neighbor mesh. This is because the communication patterns of the two applications are mostly localized. Thus, topologies with more nearest-neighbor links yield more benefits than topologies with less nearest-neighbor links.

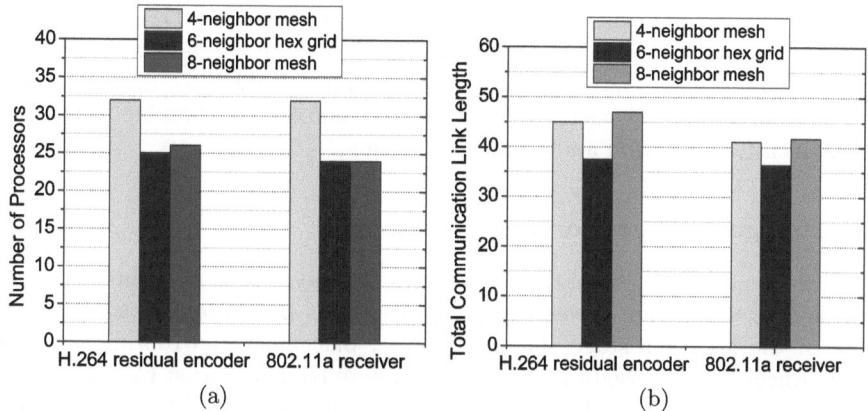

Fig. 13. The application mapping results of the 4-neighbor mesh, 6-neighbor hex grid and 8-neighbor mesh (a) the number of used processors, (b) the total communication link length

Figure 13(b) shows the total communication link length for the two applications which is calculated based on the data in Table 2 and the application mapping diagrams. The 8-neighbor mesh has longer communication length than the 4-neighbor mesh because of using more long communication links. The 6-neighbor hex grid is the most efficient topology, yielding the largest reduction (19%) in average total communication link length compared to the baseline 4-neighbor mesh.

5   Physical Design Methodology and Hexagonal Processor Tile Design

5.1   Physical Design Methodology

For performance evaluation, a small DSP processor with configurable circuit-switch interconnection is used for all physical designs. The processor contains a 16-bit datapath with a 40-bit accumulator and 560-Byte instruction and 256-Byte data memories. Each processor also contains two 128-Byte FIFOs for data buffering and synchronization between two processors.

Each set of inter-processor links are composed of 19 signals including a clock, 16-bit data and 2 flow-control signals. This processor is tailored for all topologies under test with a different number of neighboring interconnections ranging from 4 to 8. The internal switch fabrics are changed accordingly. The hardware overhead is minimal for 6-neighbor and 8-neighbor processors with only 0.7% and 2.0% hardware overhead based on the synthesis results. In order to make CMP integration simpler, four additional sets of pins are inserted into the processor netlist after synthesis and are directly connected with bypass wires for the 8-neighbor processor. This adds routing congestion in the corner for the 8-neighbor mesh topology shown in Fig. 4(c).

The processors are synthesized from Verilog with Synopsys Design Compiler and laid out with an automatic timing-driven physical design flow with Cadence SoC Encounter in 65 nm CMOS technology. Timing is checked and optimized after each step of the physical design flow: floorplan, power planning, cell placement, clock tree insertion and detailed routing.

5.2   Hexagonal Processor and CMP Design

The hexagonal-shaped tile bring challenges for physical implementation. The first challenge to design the hexagonal processor is how to create a hexagonal shape at the floorplan stage. The rectangular placement and routing blockage in SoC Encounter are used to create approximate triangle corner blockages with each rectangular blockage differs by one unit in width and height. All rect blockages are piled together to create an approximate triangle in the four corners of the rectangular floorplan as shown in Fig. 14.

A proper placement of pins can help to achieve efficient global routing and easy CMP integration. At the floorplan stage, four sets of pins are put along the diagonal edge of the corner and two set of pins are placed in the horizontal top and bottom edge. Since all macroblocks have rectangular shapes (IMEM, DMEM and two FIFOs), this presents a challenge to place the macroblocks. In this design, the macroblocks are placed along the edge and the IMEM is placed in the right corner, respectively as shown in Fig. 14.

The metal 6 and metal 7 are used to distribute power over the chip and the automatically-created power stripes can stop at the created triangle edge in the corner. The power pins are created on the top and bottom horizontal edges. When integrating the hexagonal processor together, the power nets along the triangle edge can be connected automatically or manually by simple abutment.

Once a hexagonal processor tile is laid out, a script is used to generate the RTL files of the multiprocessor. The CMP array can be synthesized with empty processor tiles inside. Another script places the hexagonal tiles with the blockage area overlap with nearest-neighbor processors along the triangle edge of each hexagonal tile. The SoC Encounter can connect all pins automatically although there are overlaps between LEF (library exchange format) files. The final GDSII files are read into Cadence icfb for design rule check (DRC). Fig. 14 shows the final layout of a hexagonal-shaped processor tile and a 6 by 6 hexagonal-tiled multiprocessor array. There are small empty spaces along the edges of the chip as described in Section 3.

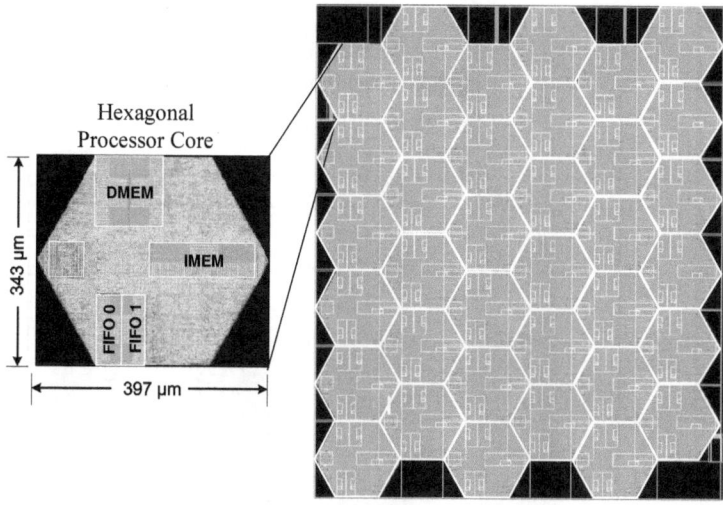

Fig. 14. Layout of a hexagonal processor and a 6x6 multiprocessor array

6   Experiment Results

6.1   Processor Implementation

All discussed topologies enable an easy integration of processors by abutment without global wires in the physical design phase. For all topologies, there is no long-distance inter-communication link across more than two processors and the processor has been pipelined in a way that the critical path is not in the interconnection links. Therefore, the maximum achievable frequency of an array is the same as an individual core, which is one of the key advantages of our proposed dense on-chip networks. Three tile types are implemented from RTL to GDSII layout. In order to be fair, all floorplans use the same power distribution design and the I/O pins and macroblocks are placed along edges reasonably depending on the topology.

In standard-cell design, the cell utilization ratio has a strong impact on the implementation result. A higher cell utilization can both save area and increase system performance if the design is routable. In order to get a minimum chip area for all tiles, we start with a relatively large tile area which results in a small cell utilization ratio. Then the tiles are repeatedly laid out while maintaining the aspect ratio and reducing the area by 5% in each iteration with minor pin and macroblock position adjustments in the floorplaning phase. Once a minimum area within 5% has been reached, the area change is reduced to 2.5%. The layout tool is pushed until it is not able to generate an error-free GDSII layout for all tiles. Our methodology results in a high cell utilization for all three tiles ranging from 81% to 83%.

Figure 15 shows the normalized implementation results of the three processor tiles in terms of area, max clock frequency, energy per operation and clock skew.

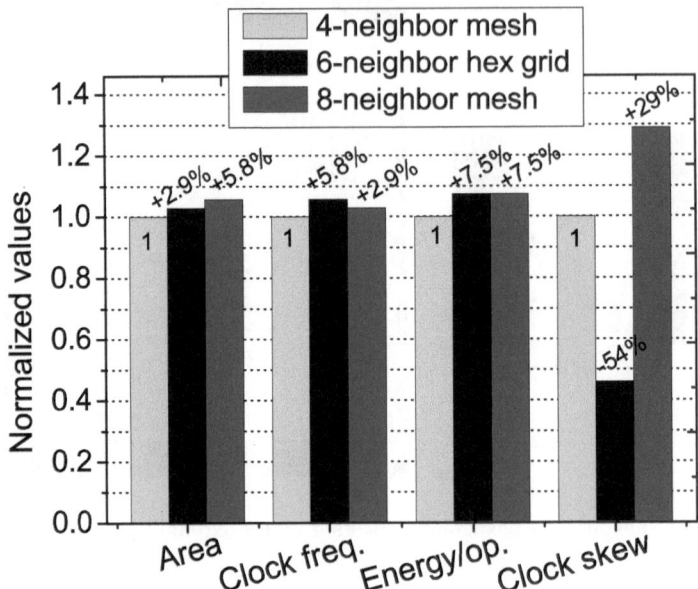

Fig. 15. Comparison of key metrics of the three optimized processor tile layout: normalized area, maximum clock frequency, energy per operation, and clock skew

The baseline 4-neighbor rectangular tile has the smallest area and the highest cell utilization of 83%. Compared with the baseline 4-neighbor rectangular tile, an area increase of 2.9% and 5.9% are required for the 6-neighbor hexagonal-shaped tile and the 8-neighbor rectangular tile, respectively. Both designs have a cell utilization of 81%.

Figure 15 also depicts the normalized maximum clock frequency relative to the baseline 4-neighbor rect tile which can operate at a maximum of 1065 MHz at 1.3 V. Due to an increase of area, the 8-neighbor rect tile can operate at 2.9% higher frequency than the 4-neighbor rect tile. The 6-neighbor hexagonal-shaped tile has noticeably higher frequencies than baseline 4-neighbor rect tile, which achieves a frequency increase of 5.8%.

Figure 15 shows the energy per operation for all tiles, which is estimated based on a 20% activity factor for all internal nodes. Both the 6-neighbor hex tile and 8-neighbor rect tile have a higher energy per operation (7.5%) because of the extra circuits for interconnections.

As for clock skew, the 8-neighbor rect tile shows a 29% higher clock skew probably because routing congestion in the corners affects the clock tree synthesis. The more circular-like shape helps the layout tool for a clock tree insertion and the hexagonal-shaped tile achieves the lowest clock skew with a reduction of 54% compared to the baseline 4-neighbor rect tile.

6.2   Application Area and Power

The actual application area depends on the number of used processors and the processor tile sizes. Fig. 16 shows the normalized application area of the H.264 residual encoder and the 802.11a baseband receiver for all three topologies. The average application area reductions are 21% and 18% for the 6-neighbor hex grid topology and the 8-neighbor mesh topology, respectively. Corresponding to the largest reduction of the number of used processors, 6-neighbor hex grid topology achieves the largest application area reduction.

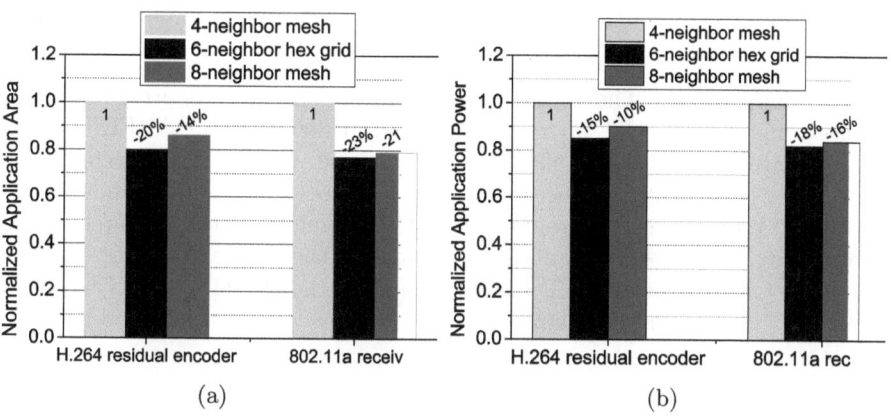

Fig. 16. The final mapping results of the H.264 residual encoder capable of HD 1080p encoding at 30 fps and 802.11a baseband receiver in 54 Mbps mode (a) normalized application area, and (b) normalized power consumption

Since tightly-tiled architecture does not have global long wires, the total application power depends on the number of used processors and the computational workload for each processor tile. In order to meet the throughput requirement for the two mapped applications, processors need to run at 959 MHz at a supply voltage of 1.15 V for H.264 residual encoder and 594 MHz at a supply voltage of 0.92 V for 802.11a baseband receiver. Based on the processor power consumption numbers, application mapping diagrams and the required clock frequencies and supply voltages for processors, Fig. 16(b) shows the normalized estimated average power consumption of the H.264 residual encoder (processing 1080p video at 30 fps) and the 802.11a baseband receiver (54 Mbps mode) for all three topologies. Compared to 4-neighbor mesh topology, the average application power reductions are 17% and 13% for the 6-neighbor hex grid and the 8-neighbor mesh topology, respectively. The 6-neighbor hex grid is the most power-efficient topology among all three topologies.

7   Conclusion

This paper presents two low area overhead and low design complexity topologies other than the commonly-used 2D mesh for tiled many-core architecture. The proposed topologies include one 6-neighbor topology which uses novel hexagonal-shaped processor tiles. This work demonstrates the feasibility of using commonly available commercial CAD tools to implement CMPs with hexagonal processor tiles. Compared to 4-neighbor 2D mesh, the proposed 6-neighbor hex grid topology has little performance and energy penalties and small area overhead while providing much more effective inter-processor interconnect to reduce application area, power consumption and total communication link lengths.

Acknowledgments. The authors gratefully acknowledge support from ST Microelectronics, Intel, UC Micro, NSF Grant 0430090 and CAREER Award 0546907, SRC GRC Grant 1598, CSR Grant 1659, Intellasys, S Machines and the support of the C2S2 Focus Center, one of six research centers funded under the Focus Center Research Program (FCRP), a Semiconductor Research Corporation entity. The authors thank Dean Truong for the chip layout assistance and Anh Tran for providing the 2D mesh mapping of the 802.11a WLAN baseband receiver, P. Cogez, and E. Flamand.

References

1. Ho, R., Mai, K., Horowitz, M.: The future of wires. Proc. of IEEE 89, 490–504 (2001)
2. Neeb, C., Wehn, N.: Designing efficient irregular networks for heterogeneous systems-on-chip. In: 9th EUROMICRO Conference on Digital System Design: Architectures, Methods and Tools (DSD 2006), pp. 665–672 (2006)
3. Leary, G., Srinivasan, K., Mehta, K., Chatha, K.: Design of network-on-chip architectures with a genetic algorithm-based technique. IEEE Transactions on Very Large Scale Integration (VLSI) Systems 17(5), 674–687 (2009)
4. Pande, P.P., Grecu, C., Jones, M., Ivanov, A., Saleh, R.: Effect of traffic localization on energy dissipation in NoC-based interconnect. In: Proc. IEEE Int. Symp. Circuits and Systems (ISCAS), pp. 1774–1777 (2005)
5. Taylor, M., et al.: A 16-issue multiple-program-counter microprocessor with point-to-point scalar operand network. In: IEEE International Solid-State Circuits Conference (ISSCC), pp. 170–171 (February 2003)
6. Yu, Z., Meeuwsen, M., Apperson, R., Sattari, O., Lai, M., Webb, J., Work, E., Truong, D., Mohsenin, T., Baas, B.: AsAP: An asynchronous array of simple processors. IEEE Journal of Solid-State Circuits 43(3), 695–705 (2008)
7. Bell, S., et al.: TILE64 processor: A 64-core soc with mesh interconnect. In: IEEE International Solid-State Circuits Conference (ISSCC), pp. 88–89 (February 2008)
8. Truong, D.N., Cheng, W.H., Mohsenin, T., Yu, Z., Jacobson, A.T., Landge, G., Meeuwsen, M.J., Tran, A.T., Xiao, Z., Work, E.W., Webb, J.W., Mejia, P.V., Baas, B.M.: A 167-processor computational platform in 65 nm CMOS. IEEE Journal of Solid-State Circuits 44(4), 1130–1144 (2009)

9. Howard, J., Dighe, S., Vangal, S., Ruhl, G., Borkar, N., Jain, S., Erraguntla, V., Konow, M., Riepen, M., Gries, M., Droege, G., Lund-Larsen, T., Steibl, S., Borkar, S., De, V., Van Der Wijngaart, R.: A 48-core ia-32 processor in 45 nm cmos using on-die message-passing and dvfs for performance and power scaling. IEEE Journal of Solid-State Circuits 46(1), 173–183 (2011)

10. Yin, A., Xu, T., Liljeberg, P., Tenhunen, H.: Explorations of honeycomb topologies for network-on-chip. In: Sixth IFIP International Conference on Network and Parallel Computing, NPC 2009, pp. 73–79 (October 2009)

11. Becker, J., Henrici, F., Trendelenburg, S., Ortmanns, M., Manoli, Y.: A continuous-time hexagonal field-programmable analog array in 0.13um CMOS with 186MHz GBW. In: IEEE International Solid-State Circuits Conference (ISSCC), pp. 70–71 (February 2008)

12. Malony, A.D.: Regular processor arrays. In: The 2nd Symposium on the Frontiers of Massively Parallel Computation, pp. 499–502 (1988)

13. Chen, M.S., Shin, K., Kandlur, D.: Addressing, routing, and broadcasting in hexagonal mesh multiprocessors. IEEE Transactions on Computers 39, 10–18 (1990)

14. Decayeux, C., Seme, D.: 3D hexagonal network: modeling, topological properties, addressing scheme, and optimal routing algorithm. IEEE Trans. on Parallel and Distributed Systems 16(9), 875–884 (2005)

15. Stojmenovic, I.: Honeycomb networks: Topological properties and communication algorithms. IEEE Transactions on Parallel and Distributed Systems 8, 1036–1042 (1997)

16. Chariete, A., Bakhouya, M., Gaber, J., Wack, M.: An approach for customizing on-chip interconnect architectures in soc design. In: 2012 International Conference on High Performance Computing and Simulation (HPCS), pp. 288–294 (July 2012)

17. Yu, Z., Baas, B.: A low-area multi-link interconnect architecture for GALS chip multiprocessors. IEEE Transactions on Very Large Scale Integration (VLSI) Systems 18(5), 750–762 (2010)

18. Xiao, Z., Baas, B.: A 1080p H.264/AVC baseline residual encoder for a fine-grained many-core system. IEEE Transaction on Circuits and Systems for Video Technology 21(7), 890–902 (2011)

19. Work, E.W.: Algorithms and software tools for mapping arbitrarily connected tasks onto an asynchronous array of simple processors. Master's thesis, University of California, Davis, CA, USA (September 2007),
http://www.ece.ucdavis.edu/vcl/pubs/theses/2007-4

20. Tosun, S., Ozturk, O., Ozen, M.: An ILP formulation for application mapping onto network-on-chips. In: International Conference on Application of Information and Communication Technologies (AICT 2009), pp. 1–5 (October 2009)

21. Kirkpatrick, S., Gelatt, C.D., Vecchi, M.P.: Optimization by simulated annealing. Science 220, 671–680 (1983)

22. Tran, A.T., Truong, D.N., Baas, B.M.: A complete real-time 802.11a baseband receiver implemented on an array of programmable processors. In: Asilomar Conference on Signals, Systems and Computers (ACSSC), pp. 165–170 (October 2008)

Fault-Tolerant Techniques to Manage Yield and Power Constraints in Network-on-Chip Interconnections

Anelise Kologeski, Caroline Concatto, Fernanda Lima Kastensmidt, and Luigi Carro

PGMICRO - PPGC - Instituto de Informática
Universidade Federal do Rio Grande do SUL (UFRGS)
Av. Bento Gonçalves, 9500 – Prédio 43412, sala 213
CEP: 91501-970 – Porto Alegre – RS - Brazil
{alkologeski,cconcato,fglima,carro}@inf.ufrgs.br

Abstract. The use of fault-tolerant mechanism is essential to ensure the correct functionality of integrated circuits after manufacturing due to the massive number of faults that may occur during the process. In this work, we propose a set of fault-tolerant techniques to cope with faulty wires in Network-on-Chip (NoC). The most appropriate technique is chosen by taking into account the number of faulty wires and their location in the NoC. The goal is to combine different techniques to reduce overheads in area, delay and power. The use of testing and diagnosis can minimize costs associated with embedded fault-tolerant mechanisms once the architecture adapts itself to work in different faulty scenarios. The proposed fault-tolerant strategy uses a lightweight adaptive routing combined with data splitting, which is able to send the data in one clock cycle. The power penalty has a low correlation with the number of faulty interconnections. Results for MPEG4 and VOPD applications running on the NoC with different faulty case-study scenarios show that the proposed techniques can tolerate many faulty interconnections with a low area, performance and power overheads.

Keywords: adaptive routing, data splitting, fault tolerance, interconnections, multiple faults, NoC.

1 Introduction

The use of embedded fault-tolerant strategies in System-on-Chip (SoC) architectures becomes crucial to improve yield and reliability, due to the huge amount of interconnections subject to the defects that comes from dimensions shrinking and aggressive transistor density. A Network-on-Chip (NoC) offers better scalability and performance than a traditional bus, and therefore it is alternative communication architecture inside of a complex System-on-Chip.

Nevertheless, according to [1], one expects up 15% of the wires faulty in recent technologies, which confirms it is necessary to consider the fault probability at the design time to ensure high yield and reliability in the devices. The use of fault-tolerant structures grows in NoC designs, due to the fact that it is almost impossible to

A. Burg et al. (Eds.): VLSI-SoC 2012, IFIP AICT 418, pp. 144–161, 2013.

manufacture integrated circuits without any defect in nanometer technologies [1]. Consequently, the use of fault-tolerant methods is crucial to allow that circuits with some amount of defects still reach the market. Therefore, fault-tolerant mechanisms in NoCs are mandatory to ensure the correct functionality, the yield and the lifetime of a chip.

The problem is that the use of several embedded fault-tolerant techniques to cope with multiple faults in links can significantly increase the overheads in area, power, energy and performance. This is because most of the techniques are applied in the critical path. In addition, they can be pre-placed even on those interconnections with no defects. To offer a flexible strategy, our proposed method combines testing and diagnosis to allows fault-tolerant techniques to be activated only in the faulty interconnections. In this way, we minimize the costs associated with the embedded fault-tolerant techniques.

The strategy presented in this work is named ATARDS -_Adaptive Technique based on Adaptive Routing and Data Splitting,_ and the strategy combines a lightweight adaptive routing (LAR) and data splitting (DS) to ensure NoC connectivity in presence of massive defects in the interconnections. The combination between two strategies allows to obtain better results when compared with traditional solution widely known in the literature. The present technique tolerates multiple faults scenarios in the interconnections (or links). Consequently, it sustains yield by keeping the connectivity in the network. If the fault-tolerant resources are configured previously due to testing and diagnosis phases, the performance and power overheads can be minimized. It is possible because the fault-tolerant techniques are enabled to operate only in faulty interconnections.

ATARDS avoids the need of additional wires in the link, and minimizes additional hardware in the critical path. The impact in performance and power is not seen in all parts of the architecture, since only faulty regions use the fault-tolerant mechanism. The experimental results with different faulty case-study NoC scenarios show the advantage of combining testing, diagnosis and ATARDS to reach better trade-offs with a high connectivity, reduced power overhead and large fault coverage.

This paper is organized as follows. Section 2 presents the fault and test models used in this work. Related work is discussed in Section 3, and we also demonstrate that our proposed strategy can fill some significant hole in the literature by cope with faulty interconnections using a merge of techniques. In Section 4, ATARDS strategy is presented with details. Results from synthesis, performance, energy and connectivity are reported in Section 5. Finally, the conclusions and ideas for future work are discussed in Section 6.

2   Fault and Test Models

The fault model provides and abstraction between the particular fault source and its manifestation in different layers of the architecture. We are mainly interested in high-level fault models in NoC. They can be in the cores (core fault model) or in the interconnections (inter-core fault model) [7]. We address permanent faults due to manufacturing. They can be modeled as shorts and open circuits. In our case, shorts will be adopted to consider a specific pattern to be addressed. A short fault occurs when a wire connects with another one. The shorts may happen among wires at the same metal layer, at the top or bottom metal layers. There are three types of short faults: OR-short, AND-short, and strong driver [19]. In a NoC, the short faults, in the wires can happen among different interconnections, from router to router or from router to core. As the amount of wires grows, the number of faults will increase exponentially [19]. Therefore, strategies that increase the number of wires are more prone to faults.

The proposed fault-tolerant techniques tolerate the inter-core faults. The inter-core fault model has been defined by faults happening among any links of the network, and it has been further classified as interlink and intralink [22]. Intralink faults happen when aggressor and victim wire are into the same link. So, they may happen isolated in links between two routers and/or in links between a core and a router. Each intralink fault is not associated with other links. Interlink faults appear when aggressor and victim are in different links. Thus, each interlink fault occurs between two different interconnections that are intersecting. Multiple defects can be any combination of intralink and interlink faults, and both can be treated by ATARDS implementation, which characterize single and multiple faulty interconnections.

Besides of permanent faults due to problems with manufacturing process, the faults also can be classified as intermittent and transient, in according to the duration. Intermittent faults occur again and again considering a certain period of time, as a periodical influence by noise or crosstalk. The transient faults usually are result from alpha particles, heavy ions and radiation, and they reach quickly the circuit affecting only memory elements changing the information by one or few clock cycles. The architecture proposed by ATARDS only copes with faulty situations located in each interconnection. As a consequence, then just permanent and intermittent faults can be tolerated by the approach if they are previously detected and diagnosed.

The capability of detecting faults in interconnections such as short circuit among channels is mandatory for yield improvement. According to [19], for full-custom layout implementations, faults between wires of distinct links are less likely, but can still be observed. So, it is mandatory to extend the fault model to include interaction faults that affect different interconnections of a NoC, like explained before with interlink faults situation.

Detection and diagnoses can be developed during manufacturing test, and off-line tests also can run during the life time of the circuit. The test, proposed in[19] detects shorts between pairs of wires (including data and control wires within a single channel or between channels) for a mesh NoC with XY routing. One has a cost-effective test, which uses a 2 x 2 NoC to deal with a fault model, which expands for a larger mesh NoC. The proposed testing approach uses Walking-One Sequence as a method to detect faults in the NoC. Furthermore, in [19] can be extended to other interaction faults in the interconnections, such as crosstalk, by adapting the test sequence. The testing approach in [10] is very similar once it uses test vectors to allow testing and diagnosis of fault interconnections. The test uses the results to configure the registers in each channel with the information about faulty links.

In [20], the authors present a method to detect defects in SoC interconnections using I_{DDT} test (analyzing the variation of the dynamic current), boundary scan and tests of delay. A *built-in self-test* (BIST) methodology for testing the inter router links of a No-Chas been proposed in [21] considering the *Maximum Aggressor Fault* (MAF) model.

In this work, we consider that the test and diagnosis is done by test vectors like the ones proposed in [19] and [10]. Analyzing the results from test, it is possible to configure the registers to inform each router of the faulty channel, and to configure the control of multiplexers used for the data splitting strategy, as will be described in detail after the related work.

3   Related Work

Related techniques to mitigate faults in the link usually based on one of the following techniques: Hamming code, parity check, retransmission, redundancy, data splitting, adaptive routing or remapping [2-10, 22]. Some of them do not need detection and diagnosis offline, because they are always detecting and correcting possible faults at run-time. Normally the authors assume a single fault scenario, which means only one faulty wire in the interconnection, in accordance to MAF model [21] or considering only one transient fault. Solutions based on error detection and correction codes (EDAC) imply extra wires for parity/check bits and extra hardware placed in the critical path, for encoding and decoding blocks in each NoC link. EDAC impacts latency and power consumption. Moreover, they can deal only with one fault per link and not multiple intralink faults. The model of a single fault per link is not valid any longer, since multiple manufacturing defects are more common to be observed and in locations close to each other, as clusters of defects in nanometer technologies [11]. Then, one requires the use of a solution to tolerate massive faults, and as a consequence, the trend is to combine different techniques to cope with, achieving high reliability with an efficient solution.

Authors in [2] propose a technique that uses Hamming Code (HC) to protect all NoC links against crosstalk, permanent and transient faults. They consider single-error correction and double-error detection (SEC/DEC) [2]. One decodes the incoming data before being stored in the FIFO, and encoded when it leaves the router. Reported results show an area overhead of 39% and a delay penalty of 32% in frequency for 180nm technology, and there is no protection for multiple faults in the link. In [3], the authors propose to combine different methods to achieve fault tolerance to crosstalk and permanent faults in NoC links. The technique uses data splitting, Hamming Code at each half of the data, and retransmission to correct crosstalk faults in the links. On top of that, triple modular redundancy (TMR) is used to protect the handshake links. The two main disadvantages of this method are the high area and power overhead, which result from the combined use of the HC and TMR, leading to a final area four times larger than the no-protected router. Besides, there is performance degradation in the network, due to the HC encoding and decoding, plus the time redundancy required for the data split technique. The latency in [3] is also increased around four times.

The technique proposed in [4] uses parity check, data splitting and retransmission of data to protect. The technique is similar to [3], but the authors propose the use of parity check to discover a faulty interconnection instead of HC. Extra bits for the parity check have been used in each half of the link to detect faulty wires, reducing the costs in relation to [3]. In the presence of faults, the erroneous half of the data is doubled and retransmitted. Due to the required retransmission in faulty cases, the performance penalty in [4] occurs only in the presence of a fault. The main disadvantage of this method is the use of extra wires and the area overhead compared to a router protected with HC only. The big problem observed in these techniques presented in [2-4] is to deal only with multiple faulty links (each one of single-fault), but not with multiple faults per link.

In [5], for a 64-core NoC with 32 bits of channel-width, the overhead in wires is about 137%, because each link protected by the Hamming code was completely duplicated (overhead of an interconnection plus wires to HC). The total area overhead is 22%, but the voltage scaling strategy has been used to reduce the power, saving 6.6% in power consumption when compared with non-protected NoC. However, only triple-error correction and quadruple-error detection are possible, considering that there is a duplicated interconnection in each link.

In [6], redundancy has been applied in some specific components inside of a 2-channel router, which means that there are two interconnections in each channel of the router to provide reliability in the links. The area overhead is between 12.5% and 15.5%, due to the number of buffers used. Results for a 64-core NoC (for link-size that occupies 5.45% of the total area) show that when there are 20 faulty wires the connectivity is around 90%, while for 100 faulty wires the connectivity becomes low, around 30%. Furthermore, the redundancy also degrades the latency.

The work in [7] proposes the use of partially faulty links when the traffic in the network is high. The main idea is to make a uniform distribution of traffic in the links. The links capacity can be split in groups of 25%, 50%, 75%, and 100% of wires, according to the faults in the link. The proposed technique has a power consumption overhead among 5% and 8% and an area overhead of 15% to 21%. However, [7] considers that all faults concentrates within the same group of wires (affecting exactly 1/4, 1/2 or 3/4 of the link), although faults can be distributed among the link. In this way, each data has a pre-defined position to be transmitted. For instance, the first bit of the data can be placed in group1-bit1, or group2-bit1, or group3-bit1 or group4-bit1. If there is a fault in each group, some bit of data always will be affected by the faulty wire in the group, and then the link must be avoided (but avoiding a link, the traffic can be damaged).

The works proposed in [23] and [24] combine mapping and adaptive routing to increase reliability in NoCs. Both works present a mapping strategy that concurrently takes into account the application core graph, the fault probability in the links and the routing. Their goal is to obtain the Pareto set of mapping configurations with customized routing functions that minimize the average latency and maximize the reliability of the application. Both proposals use the same routing algorithm (APSRA), and do not cover faults between cores and routers, just between cores. The difference between [23] and [24] is the mapping algorithm. Moreover, the proposed technique can solve the problems caused by faulty links between core and router, while in [23-24] it is not possible, reducing their efficiency to 65% in the NoC with 12 routers and cores.

The works presented in [8] and [9] use adaptive routing to avoid faulty links and faulty routers, which implies in a relative low latency overhead. However, they use virtual channels and memory tables to avoid deadlock in the network, which are normally synonymous of area overhead and excessive power consumption. Besides that, they cannot cope with faulty wires between router and cores, because there is no a redundant path to re-route the data.

In [10], the authors propose a lightweight partially adaptive (LAR) routing strategy to cope with multiple defects in each link and multiple faulty links based on minimal change in the XY path. LAR provides minimal changes in the XY path of 2-D Torus NoC, and it can cope with faults that affect up to 100% of wires in a single link, once that the faulty link can be completed avoided by using a different path to forward the packets. Consequently, virtual channels and tables are not used, and the technique in [10] has just 1% of area overhead. However, LAR cannot cope with faulty wires between a router and core, because there is no redundant path to reach its target. Besides that, LAR cannot access a router when both inputs in vertical (South and North) or horizontal (East and West) are faulty, as well as cannot leave a router with both faulty outputs, because these situations also do not allow an alternative fault-free path. Results in [10] have shown that by using only adaptive routing, 34% of faulty links would still be non-protected in a single-fault scenario. This percentage can be even higher when considering multiple faulty links and specific LAR limitations. The advantage of LAR appears in the lowest overhead in area and performance compared to the others techniques based on parity check or Hamming code, making the penalty in time and power almost imperceptible. However, LAR itself is unable to tolerate a large number of multiple faulty cases, because for many combinations of multiple faulty wires in multiple links there is no available alternative fault-free path to be used. Consequently, the combination of LAR with another fault-tolerant technique able to use faulty links in some critical cases can be used to achieve a good compromise in reliability, area, performance and power overhead.

Therefore, it is evident that we still need efficient solutions to solve the problem of multiple faults in NoC interconnections, with minimum overheads and large fault coverage. For this reason, our initial idea was to combine [10] with data splitting (DS) and re-mapping of tasks to achieve good trade-offs, as can be seen in [12]. Initially [12] has a double impact in latency for each communication through the faulty interconnection, because two clock cycles are necessary to send each data with DS strategy. However, to minimizing this impact, [12] considers re-mapping of tasks, although it could not be applied in all situations of faults, keeping sometimes the time penalty still high. In the next section, significant upgrades have been done in the proposal developed in [12]. Memory elements sensitive to the level of clock were inserted, which enables to use the data splitting in only one clock cycle. Then, the re-mapping of tasks could be removed by adding memory elements, simplifying the strategy. As the approach uses the information about the fault diagnosis together with the best fault-tolerant configuration, the proposal obtains good trade-offs in relation to traditional Hamming approach, as will be presented later.

4   The Adaptive Technique Based on Adaptive Routing and Data Splitting: ATARDS

ATARDS copes with multiple defects, interlink and intralink. ATARDS tolerates permanent faults, as shorts and open circuits, or intermittent defects such as crosstalk. ATARDS is an improvement of [12], because it does not use re-mapping of tasks and transmits a flit in two halves considering just one clock cycle. ATARDS uses latches structure to store the data and sends each half of information in different clock levels. With the new approach, the re-mapping of tasks is not necessary because a faulty interconnection does not introduce delay in the communication time (considering clock cycles). The latency in clock cycles is the same for the proposed technique and original NoC without any fault tolerance technique. The difference is in the maximum frequency for each proposal. The router frequency is limited by the hardware overhead introduced in each approach. ATARDS has lower maximum frequency when compared to the original (non-protected) router, since there is more hardware in the critical path. However, the latency (in cycles) is the same between a NoC with ATARDS or original router, but the communication time (in seconds) is different, it depends on the maximum frequency. Even with a reduction in maximum frequency, on ATARDS compared to non-protected router, it is possible to reduce the delay impact, once no extra cycle has been inserted for cases with faulty interconnections. ATARDS also does not add any extra wires in the links, as most of related work in literature does [2-6].

ATARDS has been implemented in 12-core SoCIN NoC [16] with 2D-torus topology without virtual channels. The router architecture has been implemented in VHDL, and each router can be connected to four neighboring routers with two unidirectional channel links. Each router has a local port with a processor element connected. The architecture uses packet switching and deadlock-free XY-routing. Each input channel port has a buffer with 4 slots. All routers are capable of using the lightweight adaptive routing (LAR) and data splitting (DS), however only the ones with faulty interconnections uses one of these techniques in order to minimize the overheads according to the fault case. By using test and diagnosis [10, 19-21], each router is configured with the information relative to the faulty interconnections (registers in each channel receive the information about faulty interconnections and multiplexers from DS technique receive the information about the specific faulty wires). In presence of defects, LAR technique is always the first choice, because it has minimal impact in communication time and power.

For LAR technique, the routing algorithm checks the test information before forwarding a packet. Each router is configured with the manufacturing test information about faulty-links. An additional 10-bit register is added in each router with the test results to inform if one or more of its channels are faulty. When the contemplated output channel is indicated as faulty, an alternative path replaces the original one in the header, and the packet is re-routed through the fault-free path. Each router knows the NoC size and its own position, so it can calculate the new number of steps needed for the packet in the new path. In the 2D-torus topology of size m x n, a packet has two possible routes in the same dimension: it may go k steps to one way (positive) or $m - k$ (or $n - k$) steps to the other way (negative). Though, a packet travels no more than $m - 1$ or $n - 1$ steps from source to destination when m or n is odd, or only m or n steps when they are even. As a consequence, the router dynamically changes the target address in the header in a packet when the original address intends to use a faulty link. LAR has a small impact in latency, less than 1% for the simulated cases. This little impact can be explained because on the average, the opposite path is not much larger than the original path, and for all considered cases the alternative path was not heavily congested.

However, LAR cannot cope with fault cases when there is no redundant path. For these blocking positions, the faulty channel cannot be discarded, because the connectivity needs to be sustaining, and another strategy must be used. When fault affects both input and output channels in the same direction, the router becomes inaccessible, as presented for router R6 and R11 in Figure 1. In addition, when the fault affects the channel that connects cores and routers, there is no alternative path too, as shown in Figure 1 for MED CPU and IDCT cores. The combination of LAR with another fault-tolerant technique can enable to use faulty links with a good compromise in reliability, area, performance and power overhead. So, for simplicity, one aggregates data splitting in LAR approach. Figure 1 shows the 12-core 2D-torus NoC with the MPEG4 application mapped into the system.

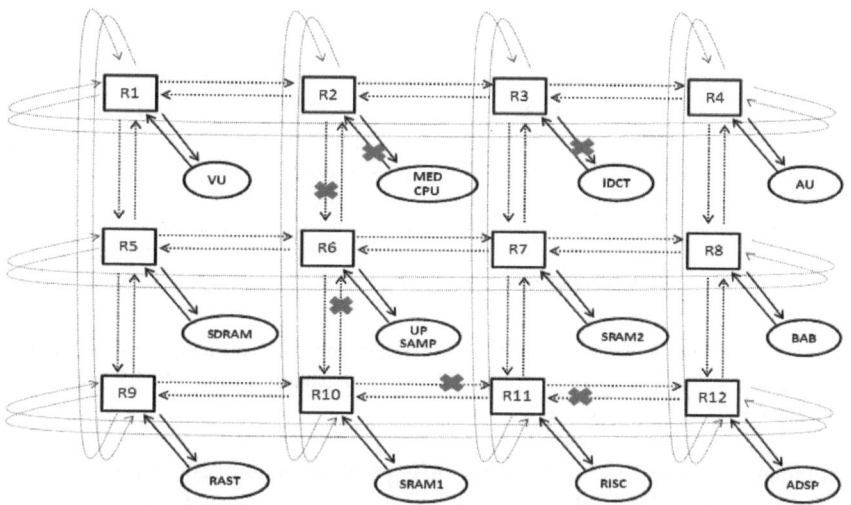

Fig. 1. MPEG4 benchmark mapped in a 12-core 2-D Torus NoC. The limitation of the LAR technique is shown by the indicated faulty links that cannot be protected by LAR mechanism.

Data splitting technique can use the partial link by selecting just the fault-free wires for each transmission. DS uses multiplexers to select the fault-free wires in each input and output channel of the router, including the local channel. The data is sent in two parts by using 50% of the interconnection in each moment. The control of each multiplexer is configured off-line based on the test results. An example of configuration can be found in Figure 2 (a) for an 8-bit link, where four faulty wires are considered. Two latches (L1 and L2) have been used to store each half of the data, because the DS technique can send the data in one cycle of the clock, at clock high and low levels. When clock is high, the first data half is transmitted through the link and it is stored in L1. At the second moment, at clock low, the second half of the data is transmitted and stored in L2. In the next cycle, both data are stored in L1 and L2 so the data can be placed together in the input buffer, as can be seen in the waveforms in

Figure 2 (b). When DS solution is not necessary, the multiplexers can be bypassed and turned off [14-15].

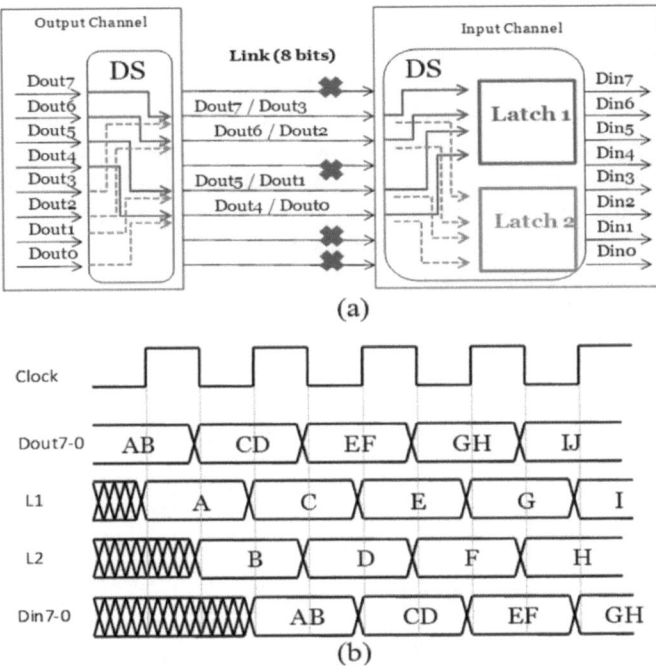

(a)

(b)

Fig. 2. DS technique implementation in the ATARDS approach: (a) an example of configuration using DS; (b) waveforms for a hypothetical communication showing the L1 and L2 latches processing the data in one clock cycle

In Figure 3, a flowchart has been used to explain the order of application of the LAR and DS strategies. Firstly the approach applies test vectors in the NoC to test and diagnose faults in the wires of the interconnection. So, after that, one tries to use LAR strategy. LAR being a possible solution to isolate the faulty links has minimum power overhead, since LAR usually introduces 1% in power overhead (see the synthesis results). LAR, needs to avoid deadlock situations: LAR can ensure a deadlock-free communication when there is only one faulty link in each interconnection's group in row and column of the NoC. A row of interconnections is, for example, all horizontal interconnections placed among Router 1, Router 2, Router 3 and Router 4 (Figure 1). A column of interconnections is a group of vertical interconnections, for example, placed among Router 1, Router 5 and Router 9 (Figure 1). When there are at least two different faulty links in a row or in a column of interconnections, it is necessary to use DS at least once. When LAR is not an option, one applies DS. DS solution will be used only when "up to X faulty wires?" is affirmative. For our approach, the X value corresponds to 50% of the wires into an interconnection. In cases when the faulty channel has more than 50% of faulty wires, the approach isolates the faulty interconnection. The flowchart needs to be repeated for each interconnection in the NoC.

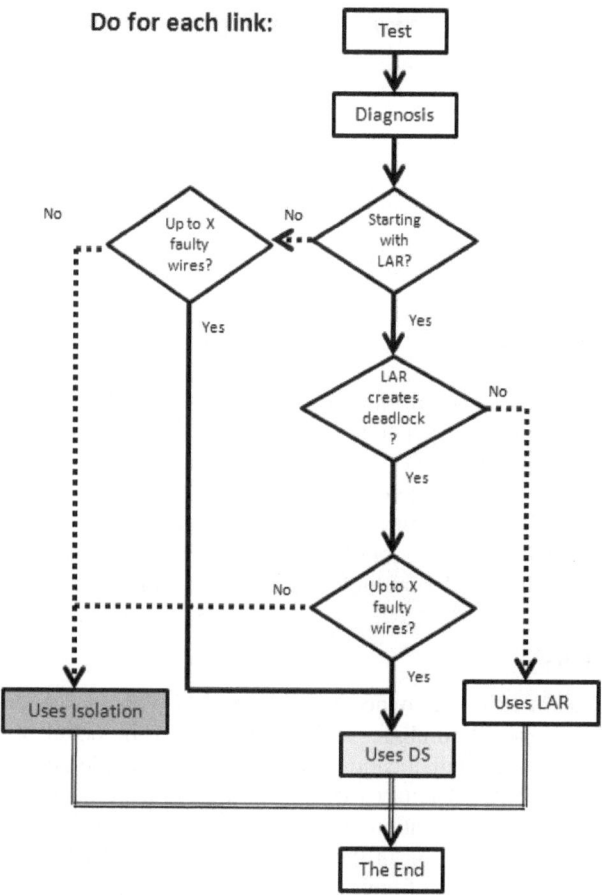

Fig. 3. Flowchart to decide what technique will be used by the ATARDS approach. The X value is 50% of the wires in the interconnection.

4.1   Fault Coverage

To compare the fault coverage, one chooses to present the following approaches for each NoC's router: non-protected, LAR, ATARDS and Hamming code (HC). To measure the fault coverage one considers two cases: acceptable number of faulty wires in each interconnection and acceptable number of faulty interconnections for each solution. Figure 4 shows a situation with only one faulty interconnection, and this interconnection can have one or more faulty wires. As LAR provides a new path when an entire link is faulty it has the best solution in that case. HC is the worst case, once it protects against only one fault per link. On the other hand, ATARDS can cope with only 50% of faulty wires and the non-tolerated strategy cannot be able to accept faults without any protection.

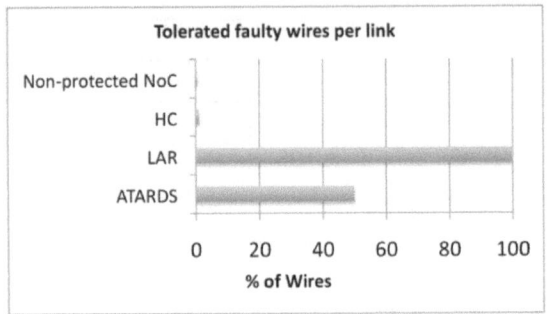

Fig. 4. Faulty tolerable number wires in each link for each strategy to sustaining connectivity

Figure 5 shows connectivity results for a scenario with multiple faulty interconnections and just one faulty wire per link. For this scenario, LAR is the weaker strategy because it accepts only one faulty interconnection in each row and each column of interconnections without causing deadlock. Thus, LAR tolerates seven faulty links of the 12-core NoC as shown in Figure 1. For multiple faulty links with single-fault, just DS and HC can protect the entire set of interconnections in the NoC with a successful rate, while DS can still consider multiple faulty wires within an interconnection.

4.2   Connectivity

ATARDS can sustain 100% of connectivity in the NoC with a large range of multiple fault combinations, once it combines techniques that can be better utilized in accordance with the type and fault location. ATARDS can completely protect the NoC when there are up to 50% of faulty wires. Multiplexers have been used to avoid the faulty wires, shifting the information into an interconnection and using the levels high and low to send the information in the same clock cycle. One expects to use LAR when there are more than 50% of faulty wires per link.

Figure 6 shows the percentage of NoC connectivity according to the faulty wires percentage in a 12-core 2-D Torus NoC with 8-bit link. We are considering the best fault distribution for each strategy. We also take into account the number of links used by the application in the NoC. The scenario regards the defect's location and the application.

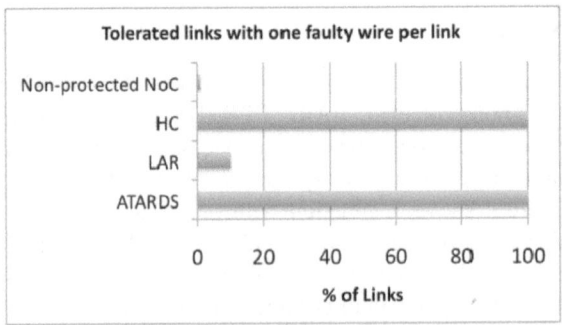

Fig. 5. Tolerated number of faulty interconnections with only one faulty wire per link in 12-core NoC

Figure 1 presents MPEG4 mapping into a NoC, for that specific mapping 24 links are not used. Wherefore, faults in these links do not affect the connectivity, and therefore we consider that faults first happen in these interconnections, for all strategies. ATARDS can sustain 100% connectivity with up to 50% of faults distributed in each interconnection because of the DS capability. For the particular case of MPEG4, it can sustain 100% connectivity even with 60% of faults (considering the best case of fault distribution).

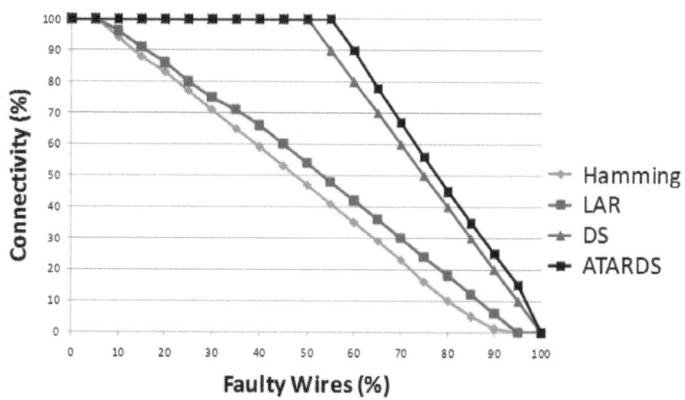

Fig. 6. NoC connectivity percentage in a generic 12-core NoC with MPEG4 benchmark

ATARDS can have better results because LAR and DS have been combined to improve the fault tolerance. Data splitting (DS) only presents good results when 50% of the wires are faulty-free in each link. LAR, by itself, can consider up seven faulty links completely faulty in a 12-core NoC to avoid deadlock situations in according to its limitations: only one fault in each row and column of the NoC can be considered, as explained in Figure 1. Hamming code has the best case of protection when there is just one faulty wire in each interconnection, that means that the efficiency is reduced as soon as possible if faults happen in more than one wire within the interconnection.

Figure 7 shows connectivity results where the number of faulty wires have been considered between 0 and 100, and the percentage calculated based on the total number of wires in the network, considering 8-bit link. The faults can be distributed in any wire of the NoC, and the worst and the best scenario compared for each approach, considering a generic case for a 4 x 3 and 8 x 8 NoC, as applied in [6]. For our approach, the best scenario happens when the faults occur in 7 specific links for a 4 x 3 NoC or 16 in an 8 x 8 NoC (ATARDS using LAR). When DS becomes necessary, the best case is when the faulty wires are at most 50% of faulty wires in each interconnection.

The best scenario for [6] occurs when the faults are completely distributed among the interconnections, because redundancy and duplication are used by the authors. The worst case for [6] is not clearly specified by the authors, but we compare with our worst case in an 8 x 8 NoC scenario (when the faults happen in more than 50% of the interconnections). For instance, when an interconnection with 8 wires has more than 4 faulty wires. Considering an 8 x 8 NoC with 100 faulty wires, ATARDS presents 6%

less of connectivity, while [6] presents almost 70% of loss. It happens because 100 faulty wires are easily tolerated by our strategy when there are many wires and inter-connections considered. When the total number of interconnections is lower, like in a 4x3 NoC case, our proposed technique shows up 30% less of connectivity, because there are few interconnections in the network and 11.57% of the total wires are faulty.

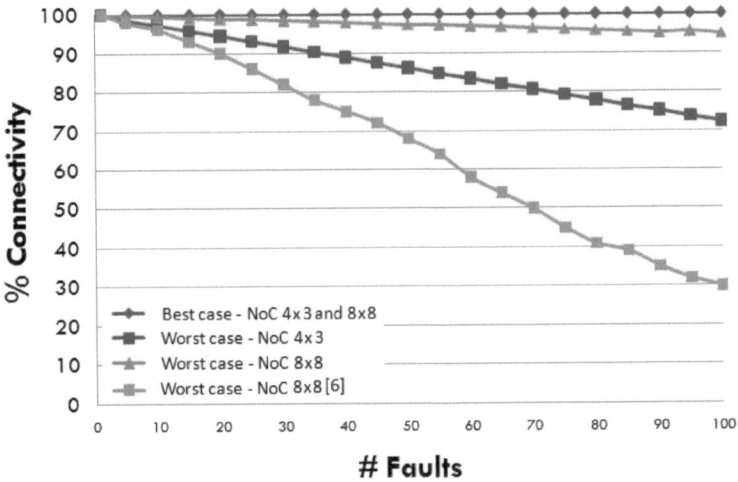

Fig. 7. Comparison of connectivity in two generic NoCs considering ATARDS and the solution proposed in [6] with 8-bit link

5   Experimental Results

ATARDS has been compared to LAR and HC technique in terms of area, perfor-mance and power. Table 1 shows the synthesis results for each router developed based on RASoC [18], with 8-bit and 32-bit links, using Synopsys Power Compiler tool with 65 nm CMOS standard cell library. Besides the data bits in the link, the SoCiN network has 2 extra wires per link to set the packet control and other 2 extra wires to do the handshake and verify the buffer availability in each channel.

ATARDS has the largest area overhead because of its configurability properties. HC presents the largest performance overhead, due to the encoding and decoding blocks with long chains of XORs. HC also incurs the largest overhead in the number of wires, once each link needs to send extra codification, using 4 and 6 extra wires for 8 and 32-bit link, respectively. Both maximum frequency and normalized frequency at 300 MHz have been considered to calculate the power results. For ATARDS, there are different types of power results, in according to the number of routers using the approach: the number of routers using DS active depends of the amount of faults present in the wires and its location. Therefore, some routers can turn off the DS solu-tion when it is not necessary in the network. Then, when ATARDS turn off the DS solution, it is running like a LAR router, and this situation is useful in the absence of faulty links to improve the power and energy results.

Table 1. Synthesis results for 65 nm technology. The total number of wires was considered for the NoC with 12-core 2-D Torus.

	Router	Area (μm^2)	Critical Path Delay (ns)	Router Power (μW)@ Max. Freq.	Router Power (μW)@ 300MHz	Total # of Wires
8 Bits	Non-protected	5360.3	1.11	334.06	111.3	864
	LAR	5260.3	1.11	338.19	112.7	864
	ATARDS (DS on)	6978.7	1.71	498.49	255.9	864
	ATARDS (DS off)	6978.7	1.71	216.64	112.7	864
	HC	5948.8	2.04	295.21	180.6	1152
32 Bits	Non-protected	13850.3	1.26	811.04	306.8	2592
	LAR	14071.3	1.26	819.12	308.4	2592
	ATARDS (DS on)	19910.4	2.09	1392.5	873.6	2592
	ATARDS (DS off)	19910.4	2.09	488.95	308.4	2592

Some power and energy results also are available for the SoCIN NoC with 12-core 2-D Torus [16]. A 4 x 3 NoC is often used by MPEG4 and VOPD benchmarks. As the behavior of these two applications is very similar, the results are in a very close range, and could be aggregated in the same value of power overhead. Figure 8 shows the best case scenario at power consumption and fault coverage for each approach (HC, ATARDS, DS and LAR). HC copes with up to 72 faulty channels, but a single-fault

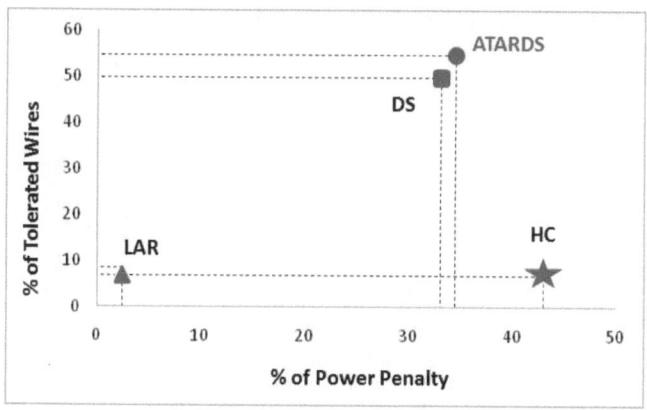

Fig. 8. The number of faulty wires tolerated by each fault-tolerant technique (considering the best fault distribution scenario) and the power penalty results, on the average, for MPEG4 and VOPD benchmarks mapped into a12-core 2-D Torus NoC. The values are very similar between 8 and 32-bit links. The results are normalized according to the power of the original non-protected NoC.

needs to be considered in each channel, showing its limitations. LAR can tolerate up to 7 completely faulty links, in a specific configuration. For multiple faulty wires and faulty links scenario HC and DS are not a solution, on the other hand, ATARDS can cope with better than just DS.

To measure the energy and power consumption of wires and routers in the NoC we considered a generic packet injection rate (1 flit/node/cycle) and a NoC size with 12-core. The power consumption in each wire has been calculated using simulations at Spice level based on the distributed π-model for the wires [17]. We assumed values between 1 mm and 1.5 mm of wire length for each link in 8 bits. The total power is the sum of the power in the router and in the wires. Figure 9 depicts the energy over-heard having the non-protected router as base, for a case considering the average between VOPD and MPEG4. The energy overhead was obtained multiplying the total power by the communication time (in seconds) at maximum frequency. HC has the higher overhead in energy because there are 4 extra wires in an 8-bit link design. LAR solution has a low impact in energy, because it is similar to non-protected router. For ATARDS there is a variable impact in energy. When there are few faulty links in the network, some ATARDS routers can avoid the DS block to improve the energy, by-passing it. For instance, when there are only 1 faulty link requiring two routers with DS on, minimizing the energy results. So, all the other routers will run with LAR on, and their DS block turned off. In the HC case, there is no variation with or without faults, and there is almost 110% of energy overhead.

Figure 10 depicts power overhead at 300 MHz. For the experiment in Figure 10, all proposed scenarios are running at the same frequency. The injection rate reduced to 25% of switching activity. ATARDS has 22% of power overhead in relation to the non-protected router, while the Hamming code has 37% of power overhead. The power results can be easily converted to energy results if you consider an equal ex-ecution time for all proposed scenarios. Latency and throughput results are not taken into account, because the frequency for the considered techniques is 300 MHz, conse-quently the traffic is the same for all situations.

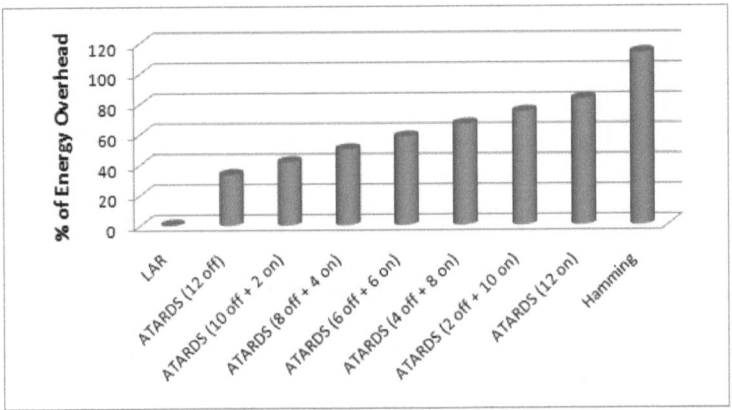

Fig. 9. The percentage overhead in energy for each configuration approach with 8-bit link design

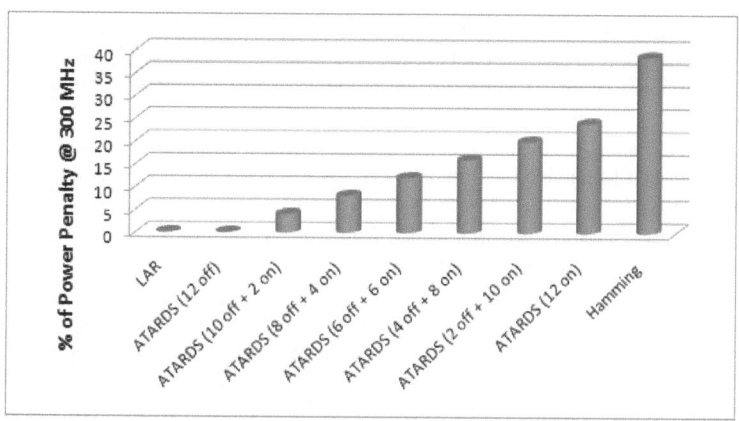

Fig. 10. Power penalty @300MHz for all strategies with 8-bit link design

6   Conclusion

An adaptive strategy for fault tolerance in NoC interconnections has been presented. The strategy named ATARDS is able to improve the yield in the presence of many faulty wires and many faulty interconnections. The technique merges LAR, DS and memory elements (latches) to decrease latency and sustain reliability. When a data uses data splitting, it is sent using only one clock cycle and no time penalty is incurred to transmit the packet through the NoC.

The DS block is used only in situations that LAR cannot cope with it. In fault-free situations, DS can be turned off to save power consumption and energy, and ATARDS runs like a LAR router. Besides that, there is no need for extra wires in ATARDS, and because that we can obtain good energy results. Moreover, ATARDS has a variable impact in power dissipation, depending on the faults location, while the HC has an excessive and constant impact. The energy can be saved when compared to well-known strategies such as Hamming code.

For MPEG4, the strategy can sustain high connectivity even when there are 60% of faulty wires, considering the 8-bit link design with the best distribution of faults. Meanwhile, HC can protect the NoC only if no more than 8.3% of the wires are faulty, considering a single-fault scenario per link.

References

1. Dehon, A., Naeimi, H.: Seven strategies for tolerating highly defective fabrication. IEEE Design & Test of Computers 22(4), 306–315 (2005)
2. Frantz, A.P., Kastensmidt, F.L., Carro, L., Cota, E.: Dependable Network-on-Chip Router Able to Simultaneously Tolerate Soft Errors and Crosstalk. In: Proceedings of 2006 International Test Conference (ITC), vol. 1, pp. 1–9 (2006)
3. Lehtonen, T., Liljeberg, P., Plosila, J.: Online Reconfigurable Self-Timed Links for Fault Tolerant NoCs. In: VLSI Design, vol. 2007, Article ID 94676, pp. 1–13 (2007)

4. Braga, M., Cota, E., Kastensmidt, F.L., Lubaszewski, M.: Efficiently using data splitting and retransmission to tolerate faults in networks-on-chip interconnects. In: Proceedings of 2010 IEEE International Symposium on Circuits and Systems (ISCAS), pp. 4101–4104 (2010)
5. Ganguly, A., Pande, P.P., Belzer, B.: Crosstalk-Aware Channel Coding Schemes for Energy Efficient and Reliable NOC Interconnects. IEEE Transactions on Very Large Scale Integration (VLSI) Systems 17(11), 1626–1639 (2009)
6. Kakoee, M.R., Bertacco, V., Benini, L.: ReliNoC: A reliable network for priority-based on-chip communication. In: Design, Automation & Test in Europe Conference & Exhibition (DATE), March 14-18, pp. 1–6 (2011)
7. Palesi, M., Kumar, S., Catania, V.: Leveraging Partially Faulty Links Usage for Enhancing Yield and Performance in Networks-on-Chip. IEEE Transactions on Computer-Aided Design of Integrated Circuits and Systems 29(3), 426–440 (2010)
8. Koibuchi, M., Matsutani, H., Amano, H., Mark Pinkston, T.: A Lightweight Fault-Tolerant Mechanism for Network-on-Chip. In: 2nd ACM/IEEE International Symposium on Networks-on-Chip, pp. 13–22 (2008)
9. Tornero, R., Sterrantino, V., Palesi, M., Ordua, J.M.: A multi-objective strategy for concurrent mapping and routing in networks on chip. In: IEEE International Symposium on Parallel & Distributed Processing, pp. 1–8 (2009)
10. Concatto, C., Almeida, P., Kastensmidt, F., Cota, E., Lubaszewski, M., Herve, M.: Improving yield of torus NoCs through fault-diagnosis-and-repair of interconnect faults. In: 15th IEEE International On-Line Testing Symposium, IOLTS 2009, June 24-26, pp. 61–66 (2009)
11. Agrawal, V.D.: Testing for Faults, Looking for Defects. In: 2011 12th Latin American Test Workshop (LATW), Keynote Talk (March 2011)
12. Kologeski, A., Concatto, C., Carro, L., Kastensmidt, F.L.: Adaptive approach to tolerate multiple faulty links in Network-on-Chip. In: 2011 12th Latin American Test Workshop (LATW), March 27-30, pp. 1–6 (2011)
13. Shih-yu, Y., Papachristou, C.A.: A method for detecting interconnect DSM defects in systems on chip. IEEE Transactions on Computer-Aided Design of Integrated Circuits and Systems 25(1), 197–204 (2006)
14. Changbo, L., Lei, H.: Distributed sleep transistor network for power reduction. IEEE Transactions on Very Large Scale Integration (VLSI) Systems 12(9), 937–946 (2004)
15. Shi, K., Howard, D.: Sleep Transistor Design and Implementation - Simple Concepts Yet Challenges To Be Optimum. In: International Symposium on VLSI Design, Automation and Test, pp. 1–4 (2006)
16. Zeferino, C.A., Susin, A.A.: SoCIN: a parametric and scalable network-on-chip. In: Proceedings of the 16th Symposium on Integrated Circuits and Systems Design, SBCCI 2003, September 8-11, pp. 169–174 (2003)
17. Sakurai, T.: Approximation of wiring delay in MOSFET LSI. IEEE Journal of Solid-State Circuits 18(4), 418–426 (1983)
18. Zeferino, C.A., Kreutz, M.E., Susin, A.A.: RASoC: a router soft-core for networks-on-chip. In: Design, Automation and Test in Europe Conference and Exhibition, vol. 3, pp. 198–203 (2004)
19. Cota, E., Kastensmidt, F.L., Cassel, M., Herve, M., Almeida, P., Meirelles, P., Amory, A., Lubaszewski, M.: A High-Fault-Coverage Approach for the Test of Data, Control and Handshake Interconnects in Mesh Networks-on-Chip. IEEE Transactions on Computers 57(9), 1202–1215 (2008)

20. Shih-Yu, Y., Papachristou, C.A., Taib-Azar, M.: Improving bus test via I_{DDT} and boundary scan. In: Proceedings of the Design Automation Conference, pp. 307–312 (2001)
21. Grecu, C., Pande, P., Ivanov, A., Saleh, R.: BIST for network-on-chip interconnect infrastructures. In: Proceedings of the 24th IEEE VLSI Test Symposium, April 30-May 4, pp. 30–35 (2006)
22. Kologeski, A., Concatto, C., Carro, L., Kastensmidt, F.L.: Improving Reliability in NoCs by Application-Specific Mapping Combined with Adaptive Fault-Tolerant Method in the Links. In: 2011 16th IEEE European Test Symposium (ETS), May 23-27, pp. 123–128 (2011)
23. Tornero, R., Sterrantino, V., Palesi, M., Orduna, J.: A multi-objective strategy for concurrent mapping and routing in networks on chip. In: IEEE International Symposium on Parallel & Distributed Processing, pp. 1–8 (2009)
24. Choudhury, A., Palermo, G., Silvano, C., Zaccaria, V.: Yield Enhancement by Robust Application-specific Mapping on Network-on-Chips. In: NoCArc 2009 - Second International Workshop on Network on-Chip Architectures, pp. 37–42 (2009)

On the Automatic Generation of Software-Based Self-Test Programs for Functional Test and Diagnosis of VLIW Processors

Davide Sabena, Luca Sterpone, and Matteo Sonza Reorda

Dipartimento di Automatica e Informatica, Politecnico di Torino, Torino, Italy
{davide.sabena,luca.sterpone,matteo.sonzareorda}@polito.it

Abstract. Software-Based Self-Test (SBST) approaches have shown to be an effective solution to detect permanent faults, both at the end of the production process, and during the operational phase. However, when Very Long Instruction Word (VLIW) processors are addressed these techniques require some optimization steps in order to properly exploit the parallelism intrinsic in these architectures. In this chapter we present a new method that, starting from previously known algorithms, automatically generates an effective test program able to still reach high fault coverage on the VLIW processor under test, while minimizing the test duration and the test code size. Moreover, using this method, a set of small SBST programs can be generated aimed at the diagnosis of the VLIW processor. Experimental results gathered on a case study show the effectiveness of the proposed approach.

Keywords: SBST, VLIW processor, Fault Simulation, Fault Diagnosis.

1   Introduction

The continuous scaling in the semiconductor fabrication process combined with the progressive growth of the integrated circuits operation frequency pushes processor cores to face more difficult testability problems. Furthermore, several phenomena such as metal migration or aging become more likely, thus increasing the occurrence of permanent faults in the generic system, in particular during the circuit operational phase. For these reasons, in order to provide high fault coverage with acceptable costs, new test solutions are being investigated and evaluated (e.g., in terms of silicon area overhead, required test infrastructure and test time).

Software-Base Self-Test (SBST) has been demonstrated to be a promising and effective approach for the test of processors and processor-based systems [1]. The SBST main idea is to generate test programs to be executed by the processor under test, able to fully stimulate the processor itself or other components belonging to the system, and to detect possible faults by looking at the produced results. The SBST technique does not require any additional hardware; therefore, the whole test cost is reduced and no performance penalty is introduced. Moreover, the SBST technique allows at-speed testing and can be easily used even for on-line test purposes. Hence,

A. Burg et al. (Eds.): VLSI-SoC 2012, IFIP AICT 418, pp. 162–180, 2013.

processor and System on Chip (SoC) testing approaches are increasingly adopting SBST techniques, often in combination with other approaches.

Correct identification of the most common defective parts in a SoC helps to characterize the technological process. The localization of a fault allows to effectively direct physical investigation of the underlying defects [2]. Moreover, a good diagnosis capability is fundamental for the devices containing self-repair skills. On the other side, it is well known that the complexity of diagnostic test generation is much higher than that of detection-oriented test generation [3]. Among the various diagnosis techniques, the Software-Based Diagnosis (SBD) methodology has turned out to be a suitable solution for processor cores embedded in SoCs [2][3].

Today, several applications demand for high performance while exposing a considerable amount of Instruction Level Parallelism (ILP), such as Digital Signal Processing [4]: among the various microprocessor architectures, Very Long Instruction Word (VLIW) processors have been demonstrated to be extremely attractive for such kinds of applications. Nowadays, several products for embedded applications adopt VLIW processors; therefore, the problem of testing them is increasingly relevant.

A major difference of VLIW processors with respect to traditional superscalar processors is the instruction format. Several VLIW instructions, named *micro-instructions*, are grouped into one large macro-instruction (also called *bundle*) where all micro-instructions within the bundle are executed in parallel computational units; each one is independent and referred to as *Computational Domain*. The operation scheduling performed by VLIW architectures is executed at compile time; therefore, the compiler is responsible for allocating the execution of each instruction to a specific Functional Unit (FU).

Due to these characteristics, VLIW processors are suitable for safety-critical systems adopted in mission-critical applications such as space, automotive or rail-transport fields which require computationally intensive functionalities combined with low power consumption. For example, the processor Tilera TILE64TM, composed of several VLIW cores, is used to efficiently perform image analysis on-board a Mars rover in support of autonomous scientific activities [5][6].

Few previously developed SBST approaches may be found in the literature in order to properly test VLIW processors against permanent faults; more in particular, part of them rely on suitable instructions belonging to the original processors instruction set to apply the test patterns previously generated by automated test pattern generation (ATPG) tools, which particularly focus on internal components [7]. These methods present some drawbacks: first of all, transforming the test patterns generated by the ATPG into test programs is not always straightforward; secondly, the resulting test programs are not optimized, especially in terms of test duration; finally, the attainable fault coverage is rarely as high as it may be required.

VLIW processors include a register file having some characteristics (in particular, the fact that it can be accessed from different domains) that make it different than the one in other processors. In [8] we focused on this component and proposed a solution, based on a SBST approach, which resulted to be quite effective.

Considering the diagnosis problem in VLIW processor, in the literature there is only a preliminary work aimed at the localization of permanent defects inside VLIW components, and the provided solution is a combination of several self-test techniques (SBST and BIST) [9].

In this chapter we focus on the generation of effective SBST test programs for VLIW processors, characterized by minimal size, minimal duration and maximum fault coverage. The proposed method starts from existing SBST test programs developed for the different FUs embedded into most processors (e.g., ALUs, adders, multipliers and memory units). Although the characteristics of FUs used within a VLIW processor are similar to those used in traditional processors, generating optimized code to effectively test these units is not a trivial task: our test generation procedure addresses the several units embedded into distinct parallel computational domains, thus taking into consideration the inherently parallel architecture of VLIW processors. Another goal of our work was the development of a general approach that could lead to the automatic generation of the test program for a VLIW processor, once the test code for testing each unit is available, and the processor configuration is known. The architecture of a VLIW processor does not include any custom hardware module, but rather a combination of common Functional Units. Our solution allows test program generation and optimization to be performed autonomously, while automatically exploiting the VLIW characteristics, without any further manual effort. The proposed method allows to generate highly optimized test programs which exploit most of the VLIW processor features and are aimed at minimizing the test time and the test program size. Besides, the method does not require the usage of any ATPG tool, since it is fully functional. Finally, without any additional effort, it is possible to exploit the test programs developed during the proposed flow to perform fault diagnosis and thus identify the faulty unit among the most relevant modules of the considered VLIW processors.

The main contribution of this chapter is the description of the first technique able to completely automate the generation of effective Software-Based Self-Testing programs for VLIW processors, while guaranteeing that the resulting programs are optimal in terms of duration and size. Exploiting this automatic method, test programs having some diagnostic properties can also be generated. The proposed method has been evaluated on a VLIW platform based on the Delft University ρ-VEX VLIW processor [10][11] which supports most of the features of industrial VLIW architectures. The results we achieved clearly demonstrate the effectiveness of our approach. Considering the generation of the optimized test programs, clock cycles have been reduced by approximately 54% with respect to the original test programs, while the size of the optimized test program decreased by approximately 58%. When the diagnosis capabilities are considered, given a generic fault in the VLIW processor under test, we are able to distinguish it uniquely in the 2.78% of the cases; moreover, in 79.15% of cases we are able to identify the faulty module containing the fault itself, while in the remaining cases we are able to narrow down the set of candidate faulty modules to 2 modules (54.52%) or to 3 modules (38.81%).

The chapter is organized as follows. Section 2 gives an overview of the VLIW architecture. Section 3 describes the related work on Software-Based Self-Test techniques specifically oriented to VLIW processors, while Section 4 explains in detail the proposed method. Experimental results on the selected case study and their analysis are presented in Section 5. Finally, conclusions and future work are described in Section 6.

2   VLIW Architecture Summary

The main characteristic of a VLIW processor is the fact that all the operations are executed by parallel *Computational Domains*, each one characterized by its own Functional Units. Besides, the scheduling is totally static, since compile tools preliminary define it at compile time. As illustrated in Fig. 1, the assembly code for a VLIW processor is drastically different from the point of view of the machine code with respect to a superscalar processor: several instructions are grouped together in a single macro-instruction (named *Bundle*) and for each instruction there are some information items that allow to assign its execution to a specific Computational Domain. Consequently, in a VLIW processor there isn't any hardware instruction scheduler, and the tasks typically performed by this component are done by the compiler. The power consumption is thus reduced and the silicon area decreases if compared to traditional superscalar processors. Furthermore, the Instruction Level Parallelism (ILP) can be adequately exploited (at least in the case of data intensive applications) since a good compiler is able to decide which instructions can be executed in parallel by checking the entire program at compile time [8].

A generic VLIW processor parametric architecture may have a variable number of functional units (FUs), so that different options, such as the number and type of functional units, the number of multi-ported registers (i.e., the size of the register file), the width of the memory buses and the type of different accessible FUs, can be modified depending on the application requirements [4].

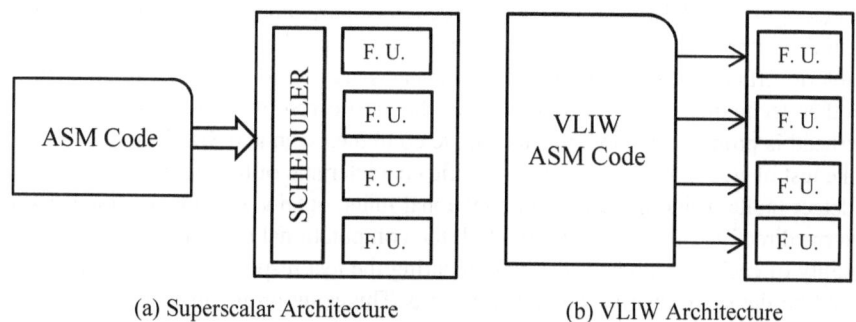

(a) Superscalar Architecture   (b) VLIW Architecture

Fig. 1. Architectural differences between a superscalar and a VLIW CPU

All the characteristics of a specific VLIW processor are grouped together and are listed in the so called *VLIW manifest*. The manifest specifies the number of computational domains, the number and type of the Functional Units embedded into each computational domain, the size and access mode of the register file and any other feature that must be taken into account when developing the code for the processor.

3   Related Work

Methodologies that require an external tester to perform the test are infeasible without the use of very expensive Automatic Test Equipments (ATEs); however the increasing gap between maximum ATE frequencies and SoC operating frequencies makes external at-speed testing problematic and expensive; at-speed testing is needed because of failures detectable only when the test is performed at the device operating frequency. Moreover, external test often involves long time and significant efforts to introduce the required hardware and may be characterized by long test application times [12]. While ATEs use external resources to perform testing task, BIST involves internal hardware resources: additional hardware and software are integrated into the circuit to allow it to perform self-testing. The usage of BIST leads to lower the cost of the complete test as well as the test time, maintaining or improving the fault coverage, at the cost of additional silicon area [8].

SBST techniques represent a special solution for on-chip testing [12], since they adopt existing processor resources and instructions to perform self-testing without any intrusiveness. The main advantage of the SBST methodology is that it uses only the processor functionality and instruction set for both test pattern application and output data evaluation, and thus does not introduce any hardware overhead in the design. However, software-based self-test methods may require very long programs to achieve high fault coverage of the device under test, and require ad-hoc techniques for generating suitable test programs [1][12]. Several papers are available in the literature related to methods for the functional self-test of processors, but only few of them refer to the test of Very Long Instruction Word (VLIW) processors [8][13][14][15].

In [8] we proposed a new SBST algorithm oriented to the test of the Register File of a generic VLIW processor; that paper highlights the particular structure of the register file belonging to a VLIW processor, that presents a particular structure since it is shared by all the computational domains of the processor; in particular, the proposed algorithm is able to efficiently test the complex cross-bar switch embedded into the component. Another technique able to obtain a good diagnostic resolution with a low hardware overhead is proposed in [14]; this technique combines scan and SBST and it is oriented to the test of VLIW processors. The specific characteristic of that approach is the ability to detect faults inside the processor functional units, obtained by loading the same test patterns directly to the test registers of all the computational domains. The proper functionality of each domain is tested by comparing the test response of all domains, which should be the same than in the fault-free case. This solution involves a hardware overhead of about 6% and requires that the processor run in self-test mode.

Similar to test approaches, several Software-Based Diagnosis (SBD) methods applied to processors have been recently developed. In [2] a new cost-effective approach is presented: the approach is based on the automatic generation of a diagnostic test set using an existing post-production test set; the authors propose to improve that set using an evolutionary method. In [9] the authors present a new diagnostic method for VLIW processors, based on scan-based BIST and SBST, aimed at a good diagnostic resolution with low hardware overhead. Software-based BIST is introduced for a fast diagnosis of the Computational Domains of the processor. This is an initial work in the field and it is based on the use of several existing self-test techniques; moreover, it is based on a specific VLIW processor and requires the introduction of several hardware test module in the considered processor.

4 The Proposed Method

In this chapter we describe a new method that allows the automatic generation of an optimized SBST program for a generic VLIW processor, once its specific configuration is known. The proposed method is composed of two main steps, denoted as *Fragmentation* and *Customization*; moreover, we propose two different flows specifically oriented to test and diagnosis, respectively. Considering the test flow, step C.1 is characterized by *Selection* and *Scheduling*; considering the diagnosis flow, step C.2 is characterized by *Classification* and *Equivalence Check* (Fig. 2); hereafter, the detailed description of each of these steps will be provided.

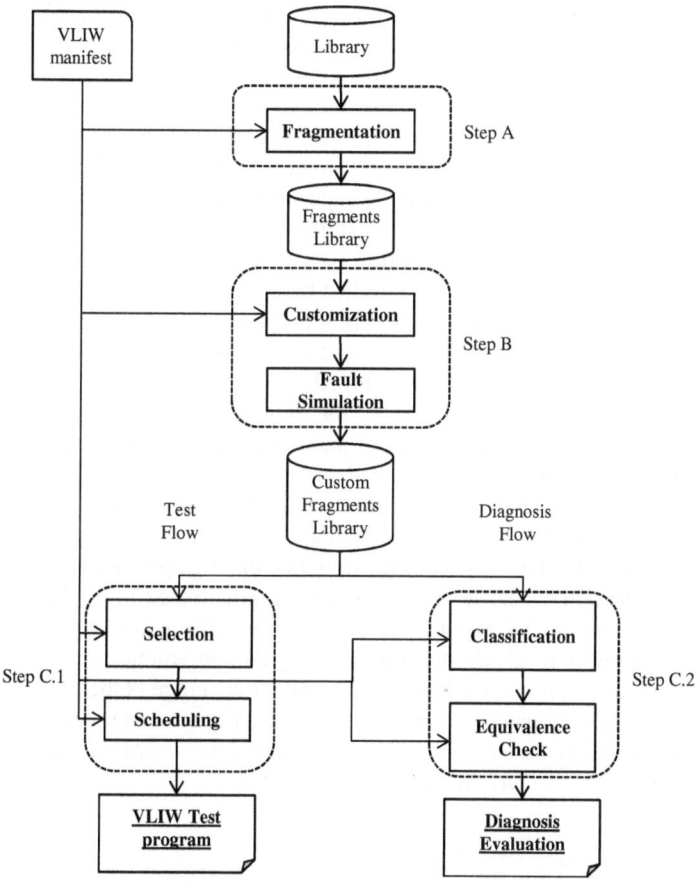

Fig. 2. The flow of the proposed test and diagnosis method

The only two requirements for the global generation flow are the manifest of the VLIW processor under test, containing all the features of the processor itself, and a library containing a set of programs able to autonomously test the different modules within the processor. The library is a collection of generic SBST programs taken from the literature [8][12][16][17][18]: it contains some functional test code able to test the most relevant Functional Units of a generic VLIW processor. The codes stored into the library are purely functional (i.e., do not require any Design for Testability feature) and are completely independent of any physical implementation of the Functional Unit they refer; these codes are described with a pseudo-code based on C language. The mapping process of these codes to the specific architecture under test is performed by the second step of the proposed method (i.e., the Customization step).

4.1   Fragmentation

The goal of the Fragmentation phase is the minimization of the number of test operations in order to generate optimized and efficient test programs. Two main tasks are performed by the Fragmentation phase: the first is the selection from the library of the test programs needed to test the VLIW processor under test, ignoring those which refer to Functional Units that are not belonging to the processor itself. The second task performed by this step is the fragmentation of each selected test program into a set of smaller pieces of code, named *Fragments*, containing few test operations and the other instructions needed to perform an independent test. The generation of a fragment is done by building it around a single instruction, and includes some preliminary instructions required to correctly perform it and to forward the results into observable locations [2][19]; the description of a Fragment is performed through some architecture-independent code. On the other hand, a test program is typically composed of a set of test operations enclosed in a loop; a series of short test programs are generated by simply separating the test operations using the Loop Unrolling technique, as shown into the pseudo-code of Fig. 3.

The code is then optimized by executing the Fragmentation phase, which exploits the fact that a VLIW processor is composed of parallel computational domains that execute several operations in parallel, as described in Section 2. Due to this feature, when a SBST program is executed with the purpose of testing a selected unit, at the same time several operations can also be executed on other parallel units. In Fig. 4 an example of this concept is shown, where it is possible to notice that by applying the SBST program for the test of the VLIW register file [8] several faults related to the Functional Units (e.g., the adders and the MEM unit) are also covered. The main idea behind test program fragmentation is to divide the original programs in atomic test units in order to effectively evaluate each one of them; multiple fault coverage is therefore avoided and the test code can be optimized in terms of test time and used resources. Once the Fragmentation phase is completed, a new library called *Fragments Library* is obtained, that contains the set of architecture-independent Fragments.

```
1.  for each cycle C of the loop L {
    1.1.    S = set of performed operations;
    1.2.    PI = input pattern applied to S into the cycle C;
    1.3.    R = expected results performing S using PI as
            input pattern;
    1.4.    GENERATE NEW FRAGMENT (PI, S, R);
2.  }
```

Fig. 3. The pseudo-code of the Fragmentation phase

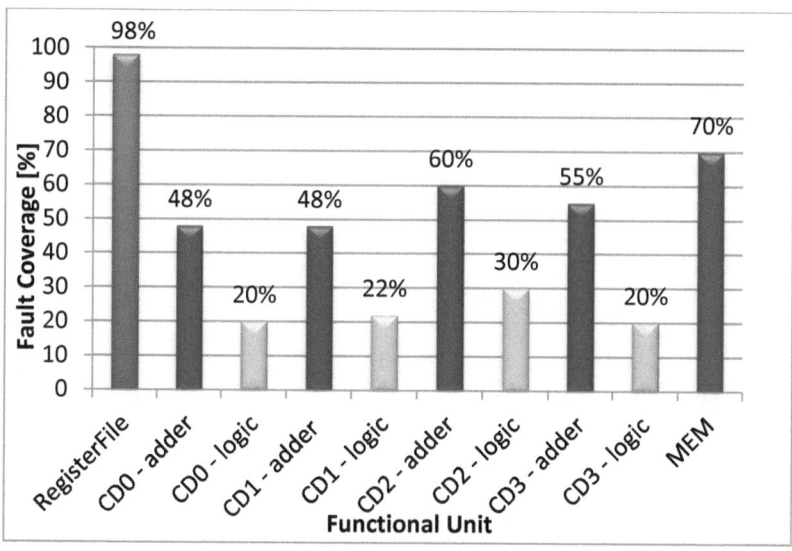

Fig. 4. The Fault Coverage of the test program for the Register File with respect to faults in the other modules of the processor

4.2   Customization

The translation of the generic architecture-independent test programs into the VLIW code is managed by the Customization step, which uses the Instruction Set Architecture (ISA) of the considered processor. In detail, starting from the VLIW manifest and from the Fragments Library, the method translates each generic Fragment into a *Custom Fragment* that can be executed by the processors under test. A Custom Fragment is defined as a set of instructions related to the ISA of the processor under test that performs several operations in order to test the addressed Functional Unit. In Table 1 an example of the Customization process is reported, where the code of a Fragment before and after the Customization phase appears. The example is based on a multiplication instruction, and the produced result is saved into the memory. As the reader can notice, at the beginning the code is a generic ISA-independent code, while after the Customization step, a VLIW code is generated, exploiting the ρ-VEX processor ISA [10][11].

The Customization phase performs two relevant tasks: the definition of the resources needed to execute the code (such as the memory area required and the registers) and the introduction of the information, inside the code, that assign the execution of an instruction to a defined VLIW Computational Domain. In Table 1, it is reported an example of this translation, where CDx is the Computational Domain in charge of executing the addressed instruction.

Table 1. Example of the translation performed by the customizer

Before Customization
R = mul (All 0's, All 0's);
Store(R , memory);
After Customization
;;----Macro-instruction 1----
CD0 : mov R1 = 0;
CD1 : mov R2 = 0;
;;----Macro-instruction 2----
CD0 : mul R3 = R1, R2;
;;----Macro-instruction 3----
CD0 : stw 4[R7] = R3;   //R7 is the stack pointer
;;----------------------------

The translation of each Fragment is performed independently from the others; furthermore, one architecture-independent Fragment can be translated into several architecture-dependent Fragments, following the features listed in the VLIW manifest, such as the type of functional units contained in each Computational Domain: for example, if in the considered VLIW processor there are 4 adder units, one for each of the 4 Computational Domains, the generic Fragment related to the test of an adder is translated into 4 architecture-dependent Fragments, one for each adder unit embedded into the Computational Domains. When the Customization phase is terminated, each architecture-dependent Fragment is fault simulated in order to compute a detailed list of faults covered by the specific test program considering all the resources of the VLIW processor. Finally, a library called *Custom Fragments Library* is obtained: it contains all the architecture-dependent Fragments used to test the processor under test and the list of faults covered by each of them. As shown in Fig. 2, the fault lists associated to each Custom Fragment are also used for the diagnosis flow, as we will explain in Section 4.4.

4.3   Selection and Scheduling

During this phase two important processes are performed: the selection of the Custom Fragments, according to the objective to be achieved, and the merge of these in order to obtain a compact and efficient test program.

Considering the Selection step, the Custom Fragments are selected by an algorithm which implements two alternative rules depending on the user requirements. The first rule is based on the selection of the minimum number of Custom Fragments that allow to reach the maximum coverage with respect to all resources of the processor under test. In this way several Custom Fragments are not selected since the faults covered by these Fragments are already covered by other fragments previously selected. The pseudo-code of this algorithm is shown in Fig. 5.

```
1. FL = Fault List of the considered processor;
2. CFL = Custom Fragments Library;
3. SFL = Selected Fragments List;
4. while ( CFL is not empty AND found) {
    4.1.   select Fragment F that allows to maximize
           the coverage of FL;
    4.2.   if (F exists){
           •   put F into SFL;
           •   remove F from CFL;
           •   found = TRUE;
    4.3.   } else
           •   found = FALSE;
5. }
```

Fig. 5. The pseudo-code of the algorithm for the selection of the Custom Fragments

The second rule is based on optimizing the number of resources used by the selected Custom Fragments. The maximal number of usable resources, in terms of registers and memory words, can be specified by the user. On the basis of these constraints, the algorithm selects the Custom Fragments that allow to reach the maximum coverage without using more resources than those specified. In this way the method is able to generate test programs depending on the final requirements: for example, if the final goal is to generate test programs for on-line testing, with the use of this algorithm we are able to generate test codes that exploit only a limited set of registers and memory words.

At the end of the Selection phase, the selected Custom Fragments enter the Scheduling phase: this process is responsible for the integration of the Custom Fragments, in order to obtain an optimized and efficient final test program. To reach this goal the scheduler optimizes and merges the codes contained into the Custom Fragments exploiting the VLIW features; in particular, it compacts the test programs aiming at maximizing the ILP of the processor. To perform the merge operation two techniques are defined and adopted; considering two or more Custom Fragments, the former is based on the exploitation of the common input pattern belonging to different instructions: in this case it is not required to define two instances of the same input data to perform the test instructions; an example of this operations is shown in Table 2, where two Custom Fragments, related to the test of the adder units embedded into the Computational Domain 0 and 1, are merged into a single test program. In this way the ILP is better exploited and the number of macro-instructions required is less than the sum of the macro-instructions of the two Fragments. The latter technique is based on the maximization of the ILP of the VLIW architecture: starting from the code of the selected Custom Fragments, the macro-instructions of these codes are merged together in order to maximize the parallel operations executed by the code.

Table 2. Example of the optimization operations performed by the scheduler

Custom Fragment A	Custom Fragment B
;;--Macro-instruction A1	;;--Macro-instruction B1
CD0 : mov R1 = 0;	CD0 : mov R1 = 0;
CD1 : mov R2 = 0;	CD1 : mov R2 = 0;
;;--Macro-instruction A2	;;--Macro-instruction B2
CD0 : add R8 = R1, R2;	CD1 : add R9 = R1, R2;
;;--Macro-instruction A3	;;--Macro-instruction B3
CD0 : stw 0[R1] = R8;	CD0 : stw 0[R1] = R9;
;;----------------------	;;----------------------

Final Test Program F
;;-- Macro-instruction F1
CD0 : mov R1 = 0;
CD1 : mov R2 = 0;
;;-- Macro-instruction F2
CD0 : add R8 = R1, R2; //tests the adder of CD0
CD1 : add R9 = R1, R2; //tests the adder of CD1
;;-- Macro-instruction F3
CD0 : stw 0[R7] = R8; //R7 is the stack pointer
;;-- Macro-instruction F4
CD0 : stw 4[R7] = R9; //R7 is the stack pointer
;;----------------------

The goal of this scheduling technique is to generate the macro-instructions of the final test program, thus reducing the whole test time. Three analysis steps are required to acquire the necessary information with respect to each Custom Fragment: the resources required by the code, such as the registers, the memory words and the Functional Units exploited; the temporal characteristics, defined as the number of clock cycles where the resources mentioned above are employed in the execution of the code; finally, the data dependences between the instructions belonging to the Custom Fragments. These pieces of information are used to create the final test program, according to the features of the VLIW processor described in the VLIW manifest. In order to do this, the scheduler uses three structures: the first is an activity frame schedule that is used to schedule the execution of the Custom Fragments into the Computational Domains: an example of this is reported in Fig. 6, where the chart representation of the activity frame schedule of the code listed in Table 2 is reported, consisting of two Custom Fragments, called A and B, each composed of three macro-instructions called A-1, A-2, A-3 and B-1, B-2, B-3, respectively. The second structure needed to create the final test program is a graph structure, where the dependences between the instructions composing the program are saved; in Fig. 7 is reported the graph structure related to the simple example shown in Table 2. Finally, the last structure is a graph containing the information about the resources, such as registers and memory word, used by the final test program for each clock cycle. At the end of this step, the final test program is generated.

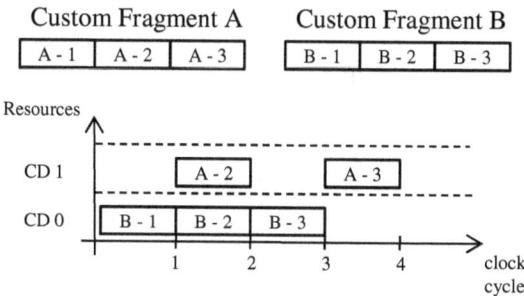

Fig. 6. The chart representation of the activity frame schedule

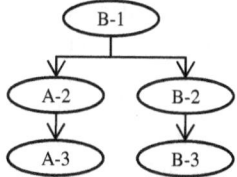

Fig. 7. The graph structure for the instruction dependence

4.4   Classification and Equivalence Check

In some situations, diagnosis is required, which means that the goal becomes the identification of the fault existing in the unit under test. For example, diagnosis is crucial
in the ramp-up phase of a new product, when the yield of the production process is
expected to grow thanks to the tuning of the process (which requires knowing where
the faults are) [20]. Another typical scenario where diagnosis is crucial is when system reconfiguration can be performed after a fault is detected, e.g., thanks to the
adoption of a programmable architecture: in this case diagnosis is crucial to identify
(once a fault is detected during the operational phase) the partition containing the
fault, so that the system can be reconfigured and the partition can be substituted by a
fault-free one [21].

Given the importance of diagnosis, we performed a preliminary analysis about the
diagnostic power of the test programs generated by our method, and we made some
considerations aimed at improving their diagnosis capabilities.

First of all, we will define the notation to be used and the steps of the diagnosis
method; then, we will report some experimental figures (in Section 5.2) about the
diagnostic capabilities of the test programs generated by the proposed method.

Notation. Let us call $F = \{f_0, f_1, ..., f_{n-1}\}$ the set of n faults that can affect the Unit
Under Test (UUT). Each of these faults causes the UUT to produce a given output
behavior b when a given sequence of input stimuli is applied; let b_i denote the output
behavior produced by fault f_i, and b_g the output behavior of the fault-free circuit.
Clearly, $b_i = b_g$ for all undetected faults f_i. In the literature (and in practice) the output

behavior can be observed (for the purpose of diagnosis) resorting to two different criteria:

- Criterion #1: the output behavior of a fault is simply the sequence of time instants in which the fault is detected. Therefore, according to this criterion $b_i = b_j$ iff the two faults f_i and f_j are detected in the same time instants.
- Criterion #2: the output behavior of a fault is the sequence of output values produced by the fault. Therefore, according to this criterion $b_i = b_j$ iff the two faults f_i and f_j always produce the same output values.

For the purpose of this paper we will consider a criterion which is a mix of criterion #1 and criterion #2. In particular, we will classify faults according to an output behavior corresponding to the set of values produced by the program at the end of its execution. Therefore, according to this criterion $b_i = b_j$ iff the two faults f_i and f_j produce the same output values in memory at the end of their execution.

A given pair of faults (f_i, f_j) is said to be *distinguished* by a given sequence of input stimuli I iff $b_i \neq b_j$. Otherwise, they are said to be *equivalent wrt I*. All faults that are equivalent wrt to a give sequence of input stimuli I are said to belong to the same *Equivalence Class wrt I*. A detected fault f_i is said to be *fully diagnosed* by a sequence of input stimuli I iff any couple of faults (f_i, f_j) including f_i is distinguished by I. Since two faults f_i, f_j can never be distinguished if they are functionally equivalent, the number of fully diagnosed faults in a circuit is typically rather low.

Several possible metrics can be adopted to measure the diagnostic capabilities of a sequence of input stimuli I [22]. A popular one is the so-called *diagnostic resolution*, or DR(I), which corresponds to the fraction of all pairs of detected faults that are distinguished by I.

When diagnosis is used in reconfigurable system for identifying the partition including the fault, the precision required is lower: in fact, the final goal in this case is to be able to distinguish all pairs of faults belonging to different partitions, while distinguishing pairs of faults belonging to the same partitions is not of interest. Hence, in this case a different definition of the diagnostic resolution can be introduced, based on a given partition of the circuit elements among P partitions. Assuming that the generic fault f_i is associated to the partition p_i, we will only consider those pairs of faults (f_i, f_j) such that $p_i \neq p_j$ and define the *partition-oriented diagnostic resolution* of a given sequence of input stimuli I, or PRDR(I), as the fraction of all pairs of detected faults belonging to different partitions that are distinguished by I.

Method. Considering the Diagnosis flow, shown in Fig. 2 Step C.2, there are two main steps necessary to acquire the diagnostic data.

First of all the fault lists associated to each Custom Fragment, and generated through fault simulation (Fig 2, Step B) are analyzed and compared: the goals of this analysis are (1) the classification of each fault, belonging to the VLIW processor under test, in the class of distinguished faults and equivalent faults, respectively, and (2) the creation of the equivalence classes, according to the notation described in the previous paragraph.

The second step is the analysis and the classification of the equivalence classes; for each of them, the classification is based on the number of partitions that have at least one fault in the considered equivalence class; the composition of the partition defines the granularity of the diagnosis and it is managed by the final user, according to the chosen diagnosis goal.

At the end of these two steps, using the obtained data and given a fault in the considered VLIW processor, we will be able to either uniquely identify it (if the fault is distinguished), or to identify the partition (one or more) containing the fault itself and the equivalent faults (if the fault has one or more equivalent).

5   Experimental Results

In this section, we present the experimental results, both for the optimized generation of the SBST program and for the diagnosis evaluation; the ρ-VEX VLIW processor has been used as a case study (Fig. 8).

The ρ-VEX is a VLIW processor released by researchers from Delft University of Technology [10][11]. Among its main features, the most important advantage is the possibility of reconfiguring the pipeline according to the user need. The pipeline, in the standard configuration, is composed of four stages: fetch, decode, execute and write-back. Following the VLIW architecture principles, the decode, execute and write-back stages are divided into four Computational Domains (CD). The fetch unit is in charge of fetching a VLIW macro-instruction from the attached instruction memory; then, it splits the considered macro-instruction into several (according to the processor configuration) micro-instructions; finally, these are passed in parallel to the decode unit. In the decoding stage two main tasks are executed: firstly, the operations are performed, and secondly the registers used as operands are fetched from the general purpose register file (the GR module of Fig. 8) and from the branch management register file (the BR module of Fig. 8). The micro-operations are then forwarded to the parallel execution units, that in this case are ALUs (1 ALU for each CD) and MULs (2 MULS, embedded in the second and in the third CD).

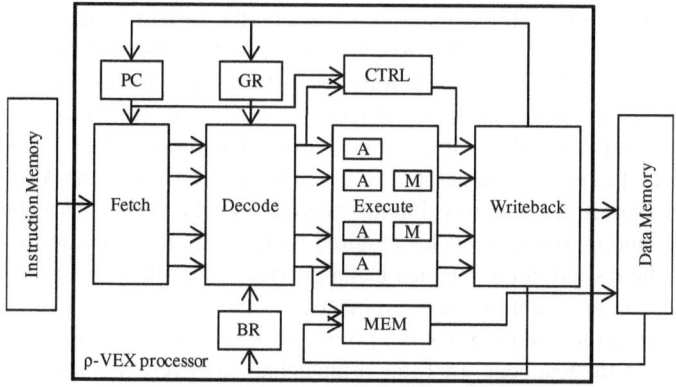

Fig. 8. The ρ-VEX VLIW processor [10][11]

In order to perform the stuck-at fault simulation experiments, we synthesized and implemented the ρ-VEX processor using a standard ASIC gate library. In total the number of faults is 387,290. The assembly code generated following the described method has been inserted into the instruction memory; then, a fault simulation experiment has been performed. Moreover, we wrote a prototypical tool (composed of about 3K lines of C++ code) implementing the proposed methods.

First of all, we have selected 6 SBST programs [8][12] [16][17][18] from the literature for testing the Functional Units embedded in the processor: each of them has been encoded in architecture-independent pseudo-code and has been inserted in the starting library. At the end of the fragmentation step we obtained a Fragments Library composed of 520 architecture-independent Fragments, while at the end of the Customization step the Custom Fragments Library was composed of 989 Custom Fragments.

5.1   Optimized SBST Program Generation Results

Using the technique for the maximum coverage with the minimum number of Fragments, 768 Custom Fragments have been selected and subjected to the scheduling step. At the end, we obtained the final test program for the test of the ρ-VEX processor: the generation time was approximately 40 hours, of which about 95% used for the fault simulation of the Custom Fragment. Computational time has been evaluated on a workstation with an Intel Xeon Processor E5450. We compared the test program generated by our approach with a test program consisting in several literature-based test programs simply queued in a unique test program, without performing any selection or scheduling steps, therefore adopting a realistic test estimation of what can be achieved with previously developed test algorithms without any optimization method. In order to fairly evaluate the two solutions, the original test programs have been applied using the loop-unrolling technique, as it is common for any VLIW application. In Table 3 we compare the obtained results.

As the reader can notice, while the coverage remains at the 98%, the number of clock cycles and the size of the test program generated with the proposed method decreased significantly. This is due to two causes: the former is that not all the Custom Fragments are chosen in the selection step; in fact the maximum coverage is reached with about 78% of the Custom Fragments. This comes from the fact that some fragments are aimed at detecting faults in some unit, which were already covered by Fragments targeted at other units. The latter is related to the scheduling step, that optimizes the code compacting the instructions, exploiting the VLIW features, and parallelizing as much as possible the execution of the Custom Fragments; consequently, the amount of clock cycles required by the final test program, is about 54% less than in the test program obtained using previously developed test programs without any selection or scheduling improvements.

It is also worth mentioning that the proposed method was able to reduce by about 58% the size of the test code. In Table 4 the achieved coverage for the relevant units of the ρ-VEX processor are reported.

Table 3. Optimized SBST program generation: obtained results

Test Program	Clock cycle [#]	Fault Coverage	Size [KB]
Original Test Programs	18,540	98.2%	3,894
Proposed method	8,447	98.2%	1,612

Table 4. Details of the achieved fault coverage

ρ-VEX Components		Faults [#]	Fault coverage
Fetch		2,156	99.2%
Decode		269,196	98.1%
Execute	4 ALU	75,554	98.3%
	2 MUL	37,244	98.6%
	MEM	1,730	97.2%
Writeback		1,420	98.1%
Total		387,290	98.2%

5.2   Diagnosis Evaluation Results

First of all we wrote a C++ program able to compare the fault lists generated by the Fault Simulation step (Section 4.2); the goal of this program is the detection of the number of distinguished faults and the classification of the undistinguished faults, i.e., the equivalent faults, in two categories: the first is composed of the faults which are equivalent and belonging to the same partition, while the second is composed of the faults belonging to different partitions. For this purpose, we divided the ρ-VEX processor in 10 partitions: the fetch unit, the decode unit, the general-purpose register file, the branch-management register file, the write-back unit, and one for each Computational Domains (i.e., 4) in which the functional units are embedded.

Then, we run this program using two different sets of fault lists: the first contains only the fault lists associated to the Custom Fragments selected by the Selection step (Fig. 2, Step C.1) of the optimized generation of the SBST program, which are 78% of the total; the second set, instead, contains the fault lists of all the Custom Fragments generated by the Customization step. In Table 5 the results of these two experiments are reported.

As it is possible to notice, the set of all fault lists (set 2) allows to increment the number of distinguished faults and the number of the equivalent faults belonging to the same partition. Consequently, considering the results of Table 5, given a fault in the ρ-VEX processor, in about 82% of the cases we are able to identify the partition affected by the fault itself.

In Table 6 the evaluation of the Equivalence Classes, generated when all the fault lists of the all Custom Fragments are considered (fault lists set 2), is shown; the purpose of this evaluation is the classification of each equivalence class, based on the number of partitions with at least one fault in the considered equivalence class. As reported in Table 6, about 93% of the equivalence classes are composed of faults belonging to the same partition. In the other cases, as reported in the graph of Fig. 9, most of the classes are composed of equivalent faults belonging to two (54.52%) or three (38.81%) different partitions.

Table 5. Faults classification: diagnosis point of view

Faults lists set	Distinguished Faults	Equivalent Faults		
		SAME partition	DIFFERENT partitions	TOTAL
1 – Optimized Test	1.13%	63.29%	35.59%	98.87%
2 - All	2.78%	79.15%	18.07%	97.22%

Table 6. Equivalence classes evaluation

Partition [#]	E.C. [#]	E.C. [%]	Faults Category
1	14,319	92.90 %	Equivalent – SAME partition
2	597	54.52 %	Equivalent – DIFFERENT partition
3	425	38.81 %	
4	42	3.84 %	
5	21	1.92 %	
6	9	0.82 %	
7	0	0.00 %	
8	1	0.09 %	
9	0	0.00 %	
10	0	0.00 %	

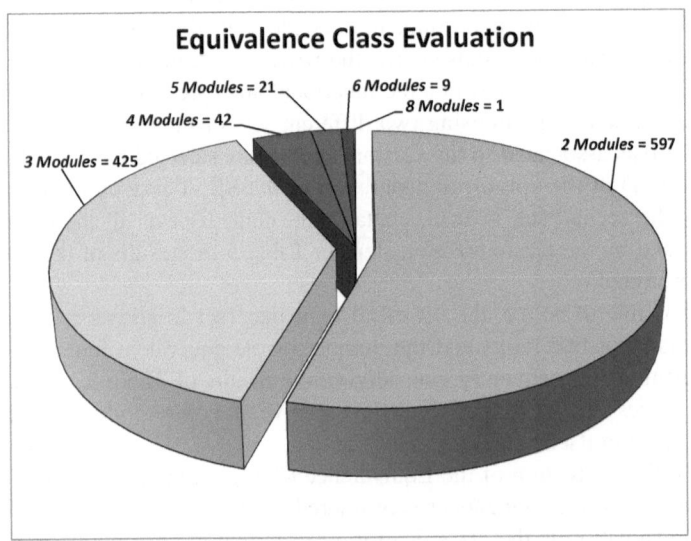

Fig. 9. The classification of the equivalent classes calculated using all the available faults lists

6   Conclusions and Future Work

In this chapter we presented the first method able to automatically generate optimized Software-Based Self-Test programs for VLIW processors. The obtained results, with respect to the selected case study, clearly demonstrate the efficiency of our method, that allows to reduce significantly both the number of clock cycles and the required memory resources with respect to the plain application of previous methods. Moreover, it is also possible to exploit the proposed method to obtain a set of small SBST programs useful for the diagnosis of the considered VLIW processor.

As future work we plan to better evaluate the performance of the proposed solution with the use of another VLIW model with different Functional Units; moreover, we plan to generate small optimized SBST programs that can be specifically used for on-line testing and able to improve the diagnosis capabilities.

References

1. Psarakis, M., Gizopoulos, D., Sanchez, E., Sonza Reorda, M.: Microprocessor software-based self-testing. IEEE Design & Test of Computers 2(3), 4–19 (2010)
2. Bernardi, P., Sànchez, E., Schillaci, M., Squillero, G., Sonza Reorda, M.: An Effective Technique for Minimizing the Cost of Processor Software-Based Diagnosis in SoCs. In: Design, Automation and Test in Europe, DATE 2006, vol. 1, pp. 1–6 (March 2006)
3. Chen, L., Dey, S.: Software-Based Diagnosis for Processors. In: Design Automation Conference 2002, pp. 259–262 (2002)
4. Fisher, J.A., Faraboschi, P., Young, C.: Embedded computing: a VLIW approach to architecture, compilers and tools. Morgan Kaufmann (2004)
5. Bornstein, B., Estlin, T., Clement, B., Springer, P.: Using a multicore processor for rover autonomous science. In: IEEE Aerospace Conference, pp. 1–9 (March 2011)
6. Tilera Corporation, "Multicore Development Environment User Guide," Doc #UG201 Release 1.2 (February 2008)
7. Beardo, M., Bruschi, F., Ferrandi, F., Sciuto, D.: An approach to functional testing of VLIW architectures. In: IEEE High-Level Design Validation and Test Workshop, pp. 29–33 (2000)
8. Sabena, D., Sonza Reorda, M., Sterpone, L.: A new SBST algorithm for testing the register file of VLIW processors. In: IEEE International Conference on Design, Automation & Test in Europe (DATE), pp. 412–417 (March 2012)
9. Ulbricht, M., Schölzer, M., Koal, T., Vierhaus, H.T.: A New Hierarchical Built-In Self-Test with On-Chip Diagnosis for VLIW Processors. In: 2011 IEEE 14th International Symposium on Design and Diagnostics of Electronic Circuits & Systems (DDECS), pp. 143–146 (April 2011)
10. Wong, S., Anjam, F., Nadeem, F.: Dynamically reconfigurable register file for a softcore VLIW processor. In: IEEE International Conference on Design, Automation and Test in Europe (DATE), pp. 962–972 (March 2010)
11. Wong, S., Van As, T., Brown, G.: ρ-VEX: a reconfigurable and extensible softcore VLIW processor. In: International Conference on ICECE Technology, pp. 369–372 (December 2010)
12. Kranitis, N., Paschalis, A., Gizopoulos, D., Xenoulis, G.: Software-based self-testing of embedded processors. IEEE Transactions on Computers 54(4), 461–475 (2005)

13. Koal, T., Vierhaus, H.T.: A software-based self-test and hardware reconfiguration solution for VLIW processors. In: IEEE Symposium on Design and Diagnostic of Electronic Circuits and Systems (DDECS), pp. 40–43 (April 2010)
14. Ulbricht, M., Scholzel, M., Koal, T., Vierhaus, H.T.: A new hierarchical built-in self-test with on-chip diagnosis for VLIW processors. In: IEEE Symposium on Design and Diagnostic of Electronic Circuits and Systems (DDECS), pp. 143–146 (April 2011)
15. Pillai, A., Zhang, W., Kagaris, D.: Detecting VLIW hard errors cost-effectively through a software-based approach. In: Advanced Information Networking and Applications Workshops, pp. 811–815 (2007)
16. Gizopoulos, D., Psarakis, M., Hatzimihail, M., Maniatakos, M., Paschalis, A., Raghunathan, A., Ravi, S.: Systematic software-based self-test for pipelined processors. IEEE Transaction on Very Large Scale Integration (VLSI) Systems 16(11), 1441–1453 (2008)
17. Paschalis, A., Gizopoulos, D., Kranitis, N., Psarakis, M., Zorian, Y.: Deterministic software-based self-testing of embedded processor cores. In: IEEE International Conference on Design, Automation and Test in Europe (DATE), pp. 92–96 (2001)
18. Kranitis, N., Gizopoulos, D., Paschalis, A., Psarakis, M.: Instruction-based self-testing of processor cores. In: IEEE VLSI Test Symposium, pp. 223–228 (2002)
19. Sanchez, E., Sonza Reorda, M., Squillero, G.: On the transformation of manufacturing test sets into on-line test sets for microprocessor. In: IEEE International Symposium on Defect and Fault Tolerance in VLSI Systems, pp. 494–502 (October 2005)
20. Bernardi, P., Sánchez, E., Schillaci, M., Squillero, G., Sonza Reorda, M.: An Effective Technique for the Automatic Generation of Diagnosis-Oriented Programs for Processor Cores. IEEE Transactions on Computer-Aided Design of Integrated Circuits and Systems 27(3), 570–574 (2008)
21. Koester, M., Luk, W.S., Hagemeyer, J., Porrmann, M., Rückert, U.: Design Optimizations for Tiled Partially Reconfigurable Systems. IEEE Transactions on Very Large Scale Integration (VLSI) Systems 19(6), 1048–1061 (2011)
22. Ryan, P.G., et al.: Fault dictionary compression and equivalence class computation for sequential circuits. In: Proc. IEEE Int. Conf. Comput.-Aided Des., pp. 508–511 (1993)

SEU-Aware Low-Power Memories
Using a Multiple Supply Voltage Array Architecture

Seokjoong Kim and Matthew R. Guthaus

University of California Santa Cruz, Santa Cruz, CA US
{seokjkim,mrg}@soe.ucsc.edu

Abstract. Electric devices should be resilient because reliability issues are increasingly problematic as technology scales down and the supply voltage is lowered. Specifically, the Soft-Error Rate (SER) increases due to the reduced feature size and the reduced charge. This paper describes an adaptive method to lower memory power using a dual V_{dd} in a column-based V_{dd} memory with Built-In Current Sensors (BICS). Using our method, we reduce the memory power by about 40% and increase the error immunity of the memory without the significant power overhead as in previous methods.

Keywords: Low-Power Memories, Single Event Upset(SEU), Soft-Error Rate (SER), Built-In Current Sensors(BICS).

1 Introduction

Single Event Upsets (SEUs) are caused by alpha particles or cosmic rays which create temporary electron-hole pairs upon collision with a silicon surface. In the past, these were common only in high-altitude (space) applications, but they are becoming more significant as process geometries and supply voltages shrink. Figure 1(a) shows the case when a particle hits the channel of a transistor. Depending on the energy and incident angle of the particle, an amount of electron hole pairs are created which can affect certain characteristics of the transistor, such as the drain current I_{ds}.

Many previous works have proposed methods to analyze transient errors induced by radiation [14, 18, 20, 27]. Basically, these methods have used a simulated pulse to emulate the spike induced by SEUs to simulate the effect at transistor/gate level as shown in Figure 1(b). In memory devices, this pulse targets the most sensitive storage node in the memory cell. The Soft-Error Rate (SER) [16,29], however, depends on the location, altitude and surrounding energy level [12] in which the circuit is operating. Researchers often use an empirical model for SER based on the critical charge Q_{crit} [4,7], but both the environmental and critical charge parameters of this model are challenging to estimate due to technology scaling and process variation. Other prior works [13] have proposed to use real measured data from radiation chambers to increase the accuracy of the prior models. This method improves the error rate accuracy, but it is costly in terms of resources and time to properly calibrate the model for the each chip designed.

To reduce the soft error rate, many previous works employ architectural techniques such as Error Correcting Codes (ECC) [5]. Error Correction Codes (ECC) add additional parity bits to original data bits to detect/correct errors. ECC can detect soft errors

A. Burg et al. (Eds.): VLSI-SoC 2012, IFIP AICT 418, pp. 181–195, 2013.
© IFIP International Federation for Information Processing 2013

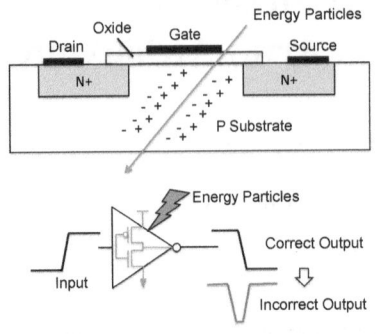
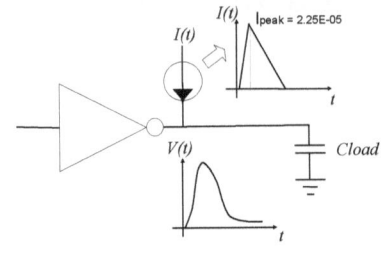

(a) Single Event Upsets (SEU) cause electron hole pairs in the transistor channel and incorrect output

(b) Gate-level soft error simulation of the impact requires temporal modeling of the charged particles

Fig. 1. Single Event Upset (SEU) example and gate-level SEU simulation methodology are used to analyze circuit robustness.

depending on the the number of parity bits. Single Error Correction Double Error Detection (SECDED) scheme is normally used for ECC due to its simple architecture, but Double Error Correction (DEC) can be implemented using more logics and gates and increases power. Also circuit sizing methods were also proposed [2]. Circuit level techniques can increase the soft error immunity using hardened memory cells. The basic idea of hardened memory cell is increasing a capacitance of stored node to increase the critical charge Q_{crit} level. This method improves the soft error tolerance but it affects to the memory performance due to the increased capacitance.

The major issue with the prior approaches is that they can't dynamically react to immediate changes in the flux energy level. Built-In Current Sensors (BICS) have been proposed that detect transient errors in real time [15, 19] so that the errors may be immediately detected and corrected by Error Correction Codes (ECC). This enables the SER to be controlled within a recoverable range while the memory operates. Although it keeps the SER within recoverable margins, the additional BICS and ECC may increase the cost and power consumption of the chip.

Dynamic Noise Margin (DNM) has been previously introduced to quantify the transient response of SRAM cells in the presence of noise [3]. DNM quantifies a memory cell's fault tolerance to a transient voltage instead of static voltage. This means that DNM can more accurately quantify the tolerance of a memory cell to realistic external noise since SEUs from alpha and neutron particles have both temporal and voltage level components. Previous researchers have proposed many different analysis methods to compute DNM [8, 22, 23, 25].

In this work, we propose a SEU-tolerant SRAM architecture using BICS to detect SEUs and then improve the dynamic noise immunity using a dual-supply Dynamic Voltage Scaling (DVS) scheme. Since most memory designs perform DVS by selecting from pre-defined V_{dd}, we propose the methods to determine the optimal supply voltage levels considering both error tolerance and power reduction based on column-based V_{dd} array architecture with BICS.

Our major contributions are as follows:

- We are the first to propose an adaptive architecture using BICS in column-based V_{dd} memory architecture.
- We are the first to quantify the optimal voltage considering power and SEU tolerance through a new Monte Carlo framework.
- We analyze the impact of peak current variation and explicitly consider the Dynamic Noise Margin (DNM) of the memory cells.
- We also show the SER improvement by increasing transistor size in memory cell.

The rest of this paper proceeds as follows: Section 2 describes the overview of our BICS architecture, Section 3 introduces our MC framework and calculates the optimal voltage levels, Section 4 describes our power model using the dual V_{dd}, Section 5 shows our experimental setup and results, and Section 6 concludes the paper.

2 Background

This section describes previous works that systematically detect transient errors in memory arrays and recent research into dual-supply voltage column based memories. The two components are integral to our approach which is presented next in Section .

2.1 Dynamic Transient Error Detection

Researchers have proposed built-in sensors to detect transient errors dynamically [15, 19,21]. Figure 2(a), for example, shows a Built-In Current Sensor (BICS) implemented alongside a representative 6T SRAM cell. The BICS connects to each column at the bottom of the array. When a particle strikes an internal node of any memory cell in the column, the voltage of the internal node fluctuates due to the electron-hole pairs and immediately decreases the virtual V_{dd} (VVDD) of the BICS. This fluctuation turns on the PMOS transistor in pull-up path of the BICS which asserts the *UPSET* signal to indicate the presence of a transient particle.

2.2 Column-Based Supply V_{dd} Array Architecture

Column-based V_{dd} memories have been recently proposed to reduce memory array power consumption [6, 26]. Figure 2(b) shows the memory array structure with each memory cell's V_{dd} is connected to the global V_{dd} in each column. Since SRAM read operations need higher V_{dd} for improved noise margins compared to write operations, a dual supply voltage saves power without performance or reliability degradation. When a column is read, the supply voltage is set to V_{high} and when a column is written, it is set to V_{low}. This approach reduces power by minimizing the supply voltage depending on the read/write operating pattern.

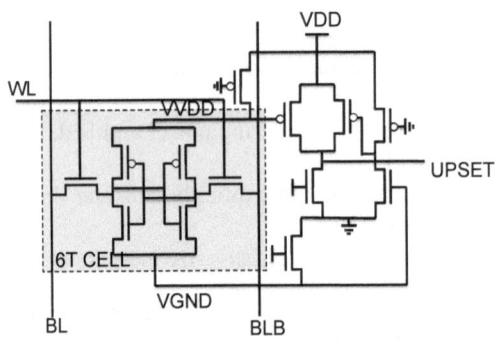
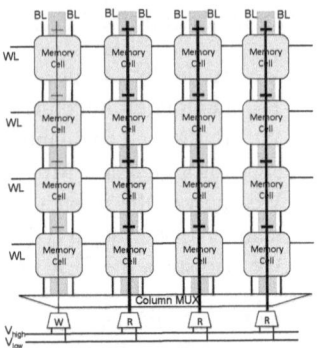

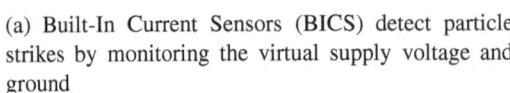

(a) Built-In Current Sensors (BICS) detect particle strikes by monitoring the virtual supply voltage and ground

(b) The column-based V_{dd} enables BICS monitoring and supply selection of individual memory columns

Fig. 2. Previous works have separately used Built-In Current Sensor (BICS) for error detection and column-based V_{dd} arrays for dynamic power savings depending on the operation (read or write)

3   Proposed Work

While column-based V_{dd} memory architectures have been used for power reduction in the previous section, our approach instead assumes the same voltage level for both operations (read and write). We alternatively combine the BICS and column-based V_{dd} array to dynamically select the minimum supply voltage to retain data values according to the present external noise conditions as shown in Figure 3. The combination of these two techniques lowers power consumption by dynamically adjusting what would otherwise be a conservative worst-case static guardband voltage while maintaining fault tolerance.

3.1   SEU-Aware Low-Power Memory Array

Our method uses the column-based V_{dd} memory architecture and BICS to detect transient errors and dynamically compensate for the noise in the memory cell using a high supply voltage. Our common-case strategy is to use a low voltage V_{dd} in normal stand-by operation and adapt with a high voltage, V_{high}, for active operation and to improve fault recovery time response. Because most memory cells spend most time in a stand-by mode, the low V_{dd} voltage efficiently reduces the stand-by leakage power. However, a low V_{dd} also reduces the robustness by directly increasing the memory cell recovery time due to transient errors. Figure 4 shows an example that illustrates how the recovery time depends on the supply voltage. The recovery time is faster with the high voltage than the low voltage.

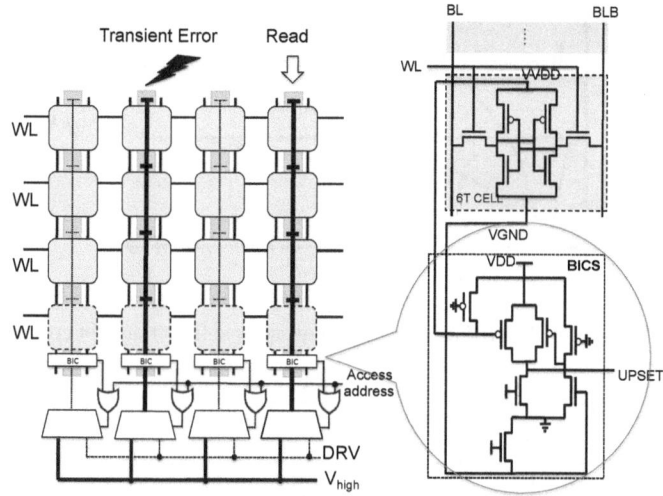

Fig. 3. Our approach uses Built-In Current Sensor (BICS) together with a Column-based V_{dd} Array to detect SEUs at a column granularity

In our approach, the supply voltage of a column is adjusted to V_{high} when a SEU is detected in memory cells in non-accessed columns. The low V_{dd} could be the Data Retention Voltage (DRV), for example, but we need a method to improve DRV robustness.

Figure 3 shows the architecture using V_{high} and DRV. V_{high} is only enabled through the supply mux when (1) the SEU occurs or (2) the column address (read/write) is addressed. To do this, we add a logical *OR* operation to the bottom of each column and connect *UPSET* signal and Column Selection (*CS*) signal as inputs. For example, if the SEU occurs in column 2 and the *CS* signal for read operation accesses column 4, only two columns are connected to V_{high}. The rest of columns are still connected to the low V_{dd}.

Our method adjusts V_{dd} of each column depending on whether a SEU is located in a column according to the BICS. This can happen in the background during idle periods. Therefore, the power consumption is reduced by only using the high supply voltage when necessary. We calculate the V_{high} supply voltages by analyzing the memory access delay constraint of a read operation and calculate the DRV using Monte-carlo SNM analysis [9]. The write operation is not directly considered, because the read operation has less noise margin and is more critical than the write operation [26].

3.2 Memory Characterization Framework

Figure 5 shows our Monte Carlo framework that is used to analyze the impact of SEUs on memory timing. It uses several configuration parameters to specify the supply voltage, memory size, device parameters, and transistor variation. Among the parameters, we consider V_{th} variation only for simplicity. It then executes two independent processes. One process performs worst case delay characterization during normal memory

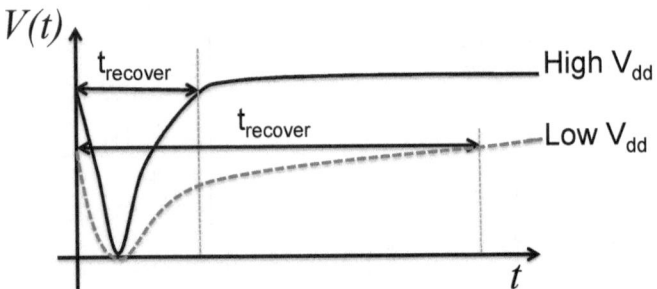

Fig. 4. Low V_{dd} reduces the cell recovery time from transient error

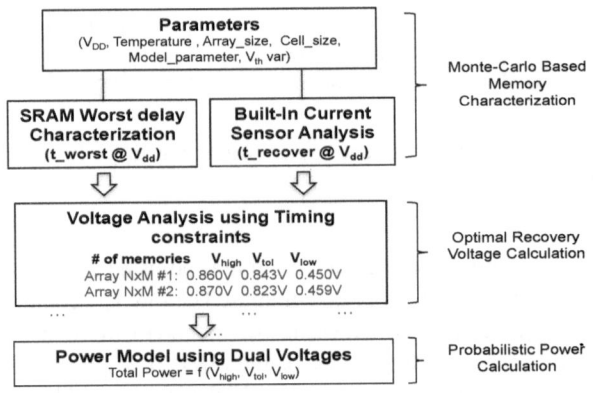

Fig. 5. A Monte Carlo framework is used to analyze the timing and power of the low and high supply voltage levels

operation while the other analyzes the recovery time when performing an access with V_{high} during a SEU. Both modules internally perform a voltage sweep to study the impact of V_{dd}.

The worst case delay is a quadratic function of the supply voltage with the coefficient depending on the array size,

$$t_{worst}(V_{dd}) = f(V_{dd}, M, N). \qquad (1)$$

Figure 6(a) shows this using simulation data (V_{dd} and array size $N \times M$). Similarly, the recovery time from a SEU using the BICS architecture is measured as the time required for a memory node voltage to fully recover (99.9% of V_{dd}) using the dual voltage. This is a function of the memory column height due to the bit-line and supply rail capacitance and the supply voltage due to the memory cell drive strength,

$$t_{recover}(V_{dd}) = f(V_{dd}, DRV, N). \qquad (2)$$

Figure 6(b) shows the recovery time $t_{recover}$ depending on column height N. As expected, large column height N increases $t_{recover}$ in both cases (I_{peak}=2.25E-05 and I_{peak}=6.25E-05) due to the linear increase in capacitance.

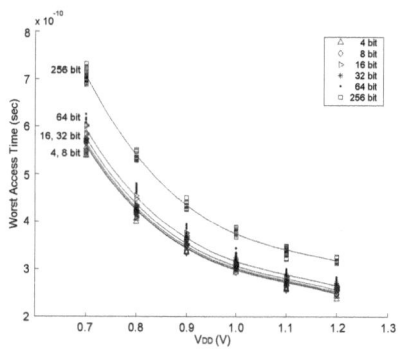

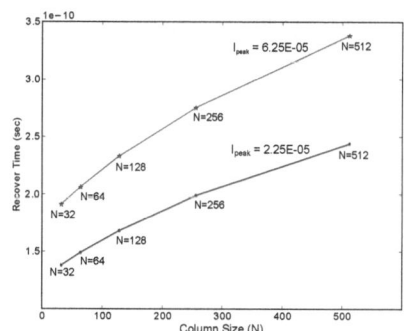

(a) Worst case delay is fit to a non-linear model for various array sizes

(b) Plot (Column size N vs. $t_{recover}$) in different I_{peak}

Fig. 6. The worst delay t_{worst} and the recovery time $t_{recover}$ are characterized independently in our Monte-Carlo Based framework

Once the memory characterization step is finished, the timing constraints are used to calculate the dual supply voltages as described in Section 3.3 and then they are used to calculate the memory power as described in Section 4.

3.3 Optimal Recovery Voltage V_{high} Analysis

V_{high} must be large enough to prevent transient errors, but it should be set at a low value to preserve power. Granting that low V_{high} can reduce the power, making V_{high} too low will reduce the transistor's driving strength so that it causes read violation errors. Our method considers the recover time $t_{recover}$ of a memory cell and the worst case delay t_{worst} without a SEU as a constraint to find a proper value of V_{high}. In our feedback architecture, the *UPSET* signal is fed to a mux to adjust the voltage to V_{high}, t_{MUX} is the time required to select the V_{dd} through the mux so that node voltage can be eventually recovered when the SEU occurs. Even after the supply voltage is adjusted to V_{high}, additional time is required to increase the voltage of memory cell internal nodes. The total recovery time is

$$t_{recover} = t_{BICS} + t_{MUX} + t_{cell}. \qquad (3)$$

Two of the sub-components (t_{BICS}, t_{MUX}) depend on the column height N while t_{cell} is largely determined by the supply voltage and cell driving strength.

The timing relation between t_{worst} and $t_{recover}$ is established as:

Criterion 1. *If a memory cell has a recovery time ($t_{recover}$) larger than the worst delay (t_{worst}), the memory cell can not recover from SEU.*

A proper V_{high} lower-bound must be calculated using two delay parameters ($t_{recover}$ and t_{worst}) at a given V_{dd}. In other words, the condition ($t_{recover} > t_{worst}$) will cause transient errors. Therefore, we can formulate the condition to avoid transient errors as:

$$t_{recover}(V_{dd}) \leq t_{worst}(V_{dd}). \qquad (4)$$

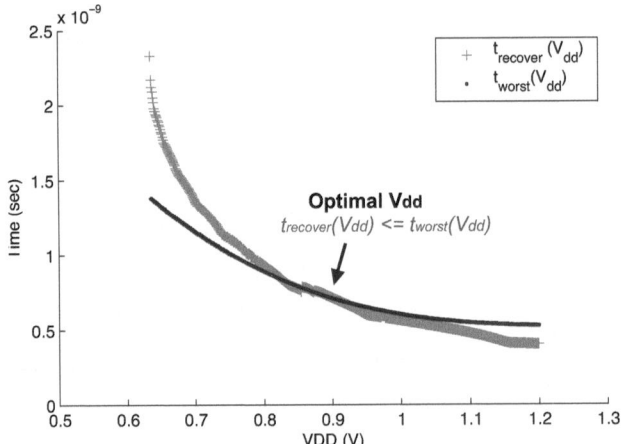

Fig. 7. Calculation of V_{high} lower-bound using t_{worst} model and $t_{recover}$ simulation with $I_{peak}=3.25E-05$ shows that the criterion is satisfied around $0.9V$ in 1024K SRAM

V_{high} is the lowest V_{dd} that satisfies Equation (4) for an given I_{peak}. We can expect that different I_{peak} can change V_{high}. This will be discussed in Section 5.1.

For example, Figure 7 shows the plot of SRAM cell $t_{recover}$ and t_{worst} at various V_{high} supply voltages. Using this plot, the V_{high} lower bound condition is satisfied near $V_{dd}=0.9V$. It is interesting to note that the quadratic coefficient of the recovery time is much less than the worst case memory. This is because the higher supply voltage enables the memory cell to recover more quickly from a SEU.

4   Power Calculation

Our architecture employs a dual voltage (V_{high} and DRV) selectively depending on the SEU occurrence and active operation frequency. This means that the V_{high} duration time differs depending on the circumstances (e.g. altitude and location) due to the flux of SEUs. This can be modeled probabilistically to estimate overall memory power.

4.1   Probabilistic Power Model using V_{high} and DRV

There are several components that must be considered to compute the power of our proposed approach. First, the column-based architecture needs an additional mux in each column to select the proper supply voltage level. Also, the BICS operates independently from read/write operations to detect transient errors. The total memory power considering these issues is estimated as

$$P_{memory} = P_{array} + P_{MUX} + P_{BICS}. \tag{5}$$

where P_{array} is the $N \times M$ array power and denoted as $P_{array}(V_{high}, DRV)$ using a cell power P_{cell} and a ratio p and (1-p). $p \in [0, 1]$ means the ratio of V_{high} duration time over total transient time. Inversely, *(1-p)* means the ratio of DRV duration over total transient time.

P_{array} is calculated using one of the following approaches: In one approach, we can see the dual V_{dd} effect in a traditional row-based array, applying V_{high} and DRV to an entire array and estimate the power as:

$$P_{array} = p \cdot \sum_{i=1}^{N} \sum_{j=1}^{M} P_{cell(i,j)}(V_{high}) + (1-p) \cdot \sum_{i=1}^{N} \sum_{j=1}^{M} P_{cell(i,j)}(DRV). \quad (6)$$

In another approach, we can apply V_{high} and DRV to columns selectively and estimate the power as:

$$P_{array} = p \cdot \{P_{col}(V_{high}) + (M-1) \cdot P_{col}(DRV)\} + (1-p)M \cdot P_{col}(DRV) \quad (7)$$

In Equation (7), P_{col} shows the power consumption of a column according to

$$P_{col} = \sum_{i=1}^{N} P_{cell}(i,j) \quad (8)$$

assuming a one bit word size. Since the memory array consists of multiple bit words, Equation (7) uses the word size W to estimate the array power according to:

$$P_{array} = p \cdot \{P_{col}(V_{high}) \cdot W + (M-W) \cdot P_{col}(DRV)\} \\ + (1-p) \cdot M \cdot P_{col}(DRV). \quad (9)$$

In order to consider the power overhead of P_{MUX} and P_{BICS}, we simulate each component using the dual voltage stimulus with probabilities p and $(1-p)$ of SEUs occurring and sum up the respective power based on the corresponding memory column size M to calculate the overall power.

5   Experimental Results

All simulations use the 45nm PTM technology models [1] with a temperature of 25°C. We assume that transistors have independent $\pm15\%/3\sigma$ variation of the nominal V_{th}. The pull-up/pull-down SRAM transistor width size ratio is 0.5 and $\frac{PR}{CR} = \frac{90nm/45nm}{180nm/45nm}$ with identical gate lengths [11, 17]. The maximum particle flux is set to the typical ground-level total neutron ($N_{flux}=56.5m^{-2}s^{-1}$ [28]) while the cross-sectional area is assumed to be $CS=0.296\mu m^2$ [24]. We generate memories ranging from 1K-256K using a memory compiler and then calculate the worst access delay based on bit-cell location using Hspice simulation. The worst case delay model t_{worst} is fit using the Matlab command *nlinfit* due to the large t_{worst} simulation time on large memory arrays.

Our results are compared to a typical guardbanded approach. The transient error tolerant voltage V_{tol} [10] is selected such that no transient errors are expected with the given maximum particle flux.

5.1 Various Peak Current I_{peak} Impact on V_{high}

Previous works modeled atomic spike pulse as an artificial current sources. The current sources are modeled as triangular model for simplicity [14, 18, 20, 27]. Without loss of generality, the energy particle injection occurs during very small time periods (less than ps). In reality, however, the induced peak current I_{peak} can be various depending on location, altitude and circumstance energy level.

We analyzed the various peak current I_{peak} ($1.315E$-$5A$ to $3.215E$-$5A$) impact on V_{high}. Figure 8 shows that I_{peak} has linear impact on V_{high} according to measurements on a 1K SRAM. We observed that the data can be modeled as linear equation

$$V_{high} = a \cdot I_{peak} + b \qquad (10)$$

where a and b coefficients calculated from the curve fitting. Equation (10) implies that high supply voltage V_{high} is necessary for low SER condition (when higher I_{peak} exists) to error tolerant. If circuits are supposed to operate with low power and designed to be tolerable, the situation that V_{high} exceeds the maximum voltage limit of the design at certain I_{peak} would be a problem, because transient errors still occur even though V_{high} is applied. So we applied some techniques to avoid this situation in Section 5.2.

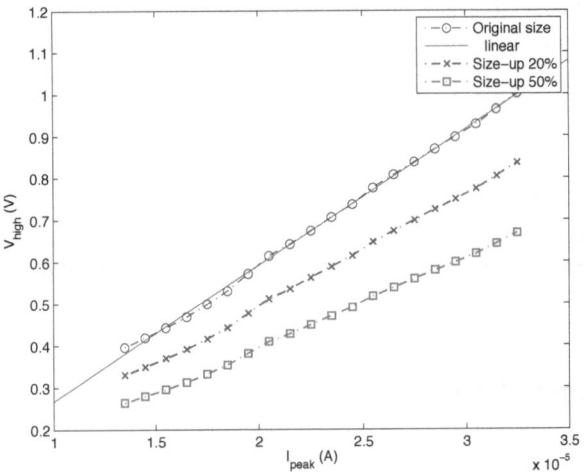

Fig. 8. Peak current's amplitude (I_{peak}) vs. V_{high} in 1K SRAM

5.2 Transistor Sizing Impact on V_{high}

We also analyzed the impact of transistor sizing on reducing V_{high} in the case that a transient error happens even with V_{high} when V_{high} may exceed the voltage budget of design. In this case, we increased width(W) and length(L) size of the SRAM cell transistors while keeping the original W/L ratio of PMOS(W/L=90nm/45nm) and NMOS(W/L=180nm/45nm) to not affect t_{worst}. We observed that transistor sizing can reduce V_{high} effectively. We increased PMOS and NMOS size by 20% and compared to to the original I_{peak} and V_{high} plot as shown in Figure 8. Sizing up transistors by 50% also show a similar trend.

For example, when the original sized SRAM cell failed at V_{high}=0.9V with given peak current I_{peak}=2.915E-5 under the maximum budget V_{dd}=0.8V, sizing up transistors by 20% can satisfy the voltage budget V_{dd}=0.8V. In the 20% sized-up SRAM plots in Figure 8, the V_{high} that satisfies the SEU tolerance under I_{peak}=2.915E-5 is about V_{high}=0.75V.

5.3   Dynamic Noise Margin (DNM) for SEU Analysis

We first analyze the Dynamic Noise Margin (DNM) during an SEU. Figure 9 shows a plot with the x-axis representing the induced peak current I_{peak} and the y-axis as recovery time $t_{recover}$. Figure 9 shows three cases using dual V_{dd} (0.4V/0.9V, 0.4V/1.2V, and 0.4V/1.5V). The vertical lines are failure points. The lines show the maximum induced noise that can be tolerated given a recovery time constraint. Using this data, we can study the DNM when dual V_{dd} can aid recovery from SEUs. and we can find the optimal V_{dd} at given I_{peak}.

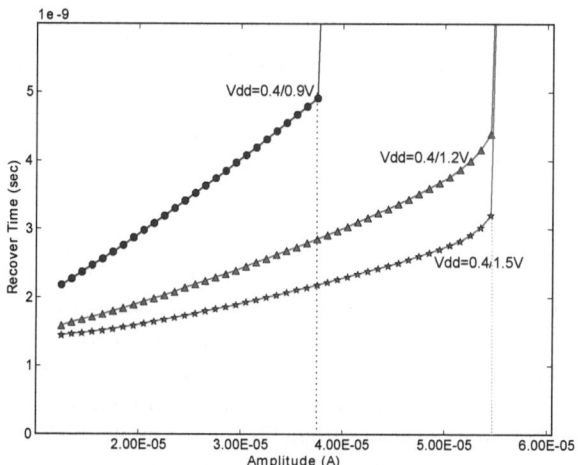

Fig. 9. Peak current's amplitude (I_{peak}) vs. $t_{recover}$ in different dual V_{dd} combinations (1K SRAM). V_{high} determines the memory tolerance to a given I_{peak} amplitude and it should be calculated to optimal V_{dd} level to reduce the power.

The DNM analysis describes whether a SEU creates a transient error or not at given I_{peak} condition. This means that we can know how dual V_{dd} schemes are tolerant to a given I_{peak}. For example, all three dual V_{dd} strategies can recover from a SEU at the condition $I_{peak} = 2.25E - 05$ although $t_{recover}$ in the case of $V_{dd} = 0.4V/0.9V$ is doubled compared to $t_{recover}$ of $V_{dd} = 0.4V/1.2V$. However, at the condition $I_{peak} = 3.75E - 05$, the $V_{dd} = 0.4V/0.9V$ case fails to recover. This means that the DNM of the memory cell determines the maximum peak noise tolerance as $I_{peak} = 3.75E - 05$

Table 1. Power Reduction Results when Radiation strikes memory Once ($p = 0.1$)

Size	SRAM with V_{tol} only	Our Proposed I (V_{high}, DRV to array) Word size = 32		Our Proposed II (V_{high}, DRV to column) Word size = 32		Word size = 8	
	Power (W)	Power (W)	Improv.(%)	Power (W)	Improv.(%)	Power (W)	Improv.(%)
1K	3.336E-06	3.430E-06	-2.81%	2.767E-06	17.06%	2.431E-06	27.14%
4K	1.321E-05	1.115E-05	15.65%	6.963E-06	47.31%	6.293E-06	52.37%
16K	5.286E-05	3.883E-05	26.54%	1.944E-05	63.23%	1.810E-05	65.76%
64K	2.114E-04	1.363E-04	35.55%	5.836E-05	72.40%	5.572E-05	73.65%
256K	8.457E-04	5.448E-04	35.58%	2.075E-04	75.46%	2.022E-04	76.09%
	Avg. Improvement(%)		16.38%		55.09%		59.00%

Table 2. Power Reduction Results when Radiation strikes memory Twice ($p = 0.2$)

Size	SRAM with V_{tol} only	Our Proposed I (V_{high}, DRV to array) Word size = 32		Our Proposed II (V_{high}, DRV to column) Word size = 32		Word size = 8	
	Power (W)	Power (W)	Improv.(%)	Power (W)	Improv.(%)	Power (W)	Improv.(%)
1K	3.336E-06	3.744E-06	-12.23%	3.216E-06	3.61%	2.543E-06	23.78%
4K	1.321E-05	1.299E-05	1.68%	7.856E-06	40.55%	6.517E-06	50.69%
16K	5.286E-05	4.744E-05	10.25%	2.122E-05	59.86%	1.854E-05	64.92%
64K	2.114E-04	1.735E-04	17.92%	6.189E-05	70.73%	5.660E-05	73.23%
256K	8.457E-04	6.975E-04	17.53%	2.146E-04	74.62%	2.040E-04	75.88%
	Avg. Improvement(%)		7.03%		49.87%		57.70%

in the case $V_{dd} = 0.4V/0.9V$. Similarly, $I_{peak} = 5.45E - 05$ is the maximum peak current tolerated with $V_{dd} = 0.4V/1.2V$ and $V_{dd} = 0.4V/1.5V$.

The DNM analysis can also be used to determine the optimal V_{dd} that can tolerate a given noise I_{peak}. As expected, higher V_{dds} enable a faster recovery time. The recovery time $t_{recover}$ of the memory cell using V_{dd}=0.4V and 1.5V's is faster than the other cases at same I_{peak}. The higher V_{dd} increases the power unnecessarily although it enables the memory cell to recover more quickly. For example, both $V_{high} = 1.2V$ and $V_{high} = 1.5V$ have the same tolerance, however, the lower voltage should be selected to save power. For this reason, the power-optimal V_{dd} should be near $V_{high} = 1.2V$ not 1.5V.

5.4   Power Reduction

We now analyze the optimal supply voltages depending on the peak current I_{peak} that a flux generates [28]. The optimal voltages are calculated as $V_{high} = 0.948V$, $V_{tol} = 0.607V$ at a flux $N_{flux} = 56.5m^{-2}s^{-1}$ and $DRV = 0.186V$. Table 1 shows the comparison of our two strategies: 1) our proposed method with V_{high} and DRV applied to the entire array (column 3-column 4), b) our proposed method with V_{high} and DRV applied to the selected columns (column 5-column 8). The baseline is a traditional SRAM with a guard-banded error-tolerant supply voltage V_{tol} (column 2).

Table 1 and Table 2 compare proposed methods when energy particles strike the memory with probabilities $p = 0.1$ and $p = 0.2$, respectively. We assume two cases since the p value is not fixed and depends on the environment where the memory operates. It can be a large number when the radiation particles strike frequently. According to Table 1, simply applying V_{high} and DRV to the entire array can reduce the power consumption by an average of 16.38% compared to an SRAM with a guard-banded supply voltage, V_{tol}. Applying V_{high} to the column with SEU and active columns selectively reduces the power consumption by an average of 55.09% compared to the guard-banded supply voltage SRAM, V_{tol}. When particles hit the memory more frequently as shown in Table 2, the power reduction decreases to 7.03% and 49.87% compared to each case in Table 1 since V_{high} is needed two times more than Table 1 to avoid errors.

We also observe that our proposed architecture increases the power consumption in the case of small memories such as 1K, due to the additional circuitry to implement the column-based V_{dd}. The additional circuitry power overwhelms the small memory array power consumption, but in large memories (4K-256K) this cost is amortized and our architecture reduces the overall power more effectively.

In both tables, we use a 32-bit word size, but we have also performed analysis with an 8-bit word size. Smaller word sizes improve the power consumption, because our architecture enables fewer columns during active read/write operations. The background recover power of memory cells in stand-by mode is not affected by the word size.

6   Conclusions

We presented a soft-error tolerant low-power memory architecture that employs BICS in column-based V_{dd} SRAM to adaptively select from dual supply voltages. We then used a Monte Carlo framework to calculate the optimal dual supply voltages and demonstrated that our architecture can significantly reduce power compared to traditional guard-banded static supply voltage architectures. On average, our architecture is able to reduce the power by an average of 39.5% without sacrificing error tolerance for an range of memory array sizes.

Acknowledgments. This work was supported in part by the National Science Foundation under grant CNS-1205493.

References

1. ASU. Predictive Technology Model (PTM), http://ptm.asu.edu
2. Bhattacharya, K., Ranganathan, N.: RADJAM: A novel approach for reduction of soft errors in logic circuits. In: VLSI Design, pp. 453–458 (January 2009)
3. Ding, L., Mazumder, P.: Dynamic noise margin: definitions and model. In: Proceedings of the 17th International Conference on VLSI Design, pp. 1001–1006 (2004)
4. Freeman, L.B.: Critical charge calculations for a bipolar SRAM array. IBM Journal of Research and Development 40, 119–129 (1996)

5. Hamming, R.W.: Error detecting and error correcting codes. Bell System Technical Journal 26(2), 147–160 (1950)
6. Hamzaoglu, F., Wang, Y., et al.: Bit cell optimizations and circuit techniques for nanoscale sram design. IEEE Design & Test of Computers 28(1), 22–31 (2011)
7. Hazucha, P., Svensson, C.: Impact of CMOS technology scaling on the atmospheric neutron soft error rate. IEEE Transactions on Nuclear Science 47(6, Part 3), 2586–2594 (2000)
8. Huang, G., Dong, W., et al.: Tracing SRAM separatrix for dynamic noise margin analysis under device mismatch. In: IEEE International Behavioral Modeling and Simulation Workshop, BMAS 2007, pp. 6–10 (September 2007)
9. Kim, S., Guthaus, M.: Leakage-aware redundancy for reliable sub-threshold memories. In: 2011 48th ACM/EDAC/IEEE Design Automation Conference (DAC), pp. 435–440 (June 2011)
10. Kim, S., Guthaus, M.: Low-power multiple-bit upset tolerant memory optimization. In: 2011 IEEE/ACM International Conference on Computer-Aided Design (ICCAD), pp. 577–581 (November 2011)
11. Kim, S., Guthaus, M.: SNM-aware power reduction and reliability improvement in 45nm SRAMs. In: 2011 IEEE/IFIP 19th International Conference on VLSI and System-on-Chip (VLSI-SoC), pp. 204–207 (October 2011)
12. Lesea, A., Drimer, S., et al.: The rosetta experiment: atmospheric soft error rate testing in differing technology fpgas. IEEE Transactions on Device and Materials Reliability 5(3), 317–328 (2005)
13. Michalak, S., Harris, K., et al.: Predicting the number of fatal soft errors in los alamos national laboratory's ASC Q supercomputer. IEEE Transactions on Device and Materials Reliability 5(3), 329–335 (2005)
14. Murley, P.C., Srinivasan, G.R.: Soft-error monte carlo modeling program, SEMM. IBM Journal of Research and Development 40(1), 109–118 (1996)
15. Neto, E., Ribeiro, I., et al.: Using bulk built-in current sensors to detect soft errors. IEEE Micro 26(5), 10–18 (2006)
16. Normand, E.: Single event upset at ground level. IEEE Transactions on Nuclear Science 43(6), 2742–2750 (1996)
17. Pavlov, A., Sachdev, M.: CMOS SRAM circuit design and parametric test in nano-scaled technologies: process-aware SRAM design. Springer (January 2008)
18. Rajaraman, R., Kim, J.S., et al.: SEAT-LA: A soft error analysis tool for combinational logic. In: VLSI Design (2006)
19. Reviriego, P., Maestro, J.A., et al.: Reliability analysis of memories protected with BICS and a per-word parity bit. ACM Trans. Des. Autom. Electron. Syst. 15, 18:1–18:15 (2010)
20. Shivakumar, P., Kistler, M., et al.: Modeling the effect of technology trends on the soft error rate of combinational logic. In: Proceedings of International Conference on Dependable Systems and Networks (DSN), pp. 389–398 (2002)
21. Vargas, F., Nicolaidis, M.: SEU-tolerant SRAM design based on current monitoring. In: Proceedings of the 24th International Symposium on Fault-Tolerant Computing (FTCS), pp. 106–115 (June 1994)
22. Vatajelu, E., Pau, G., et al.: Transient noise failures in SRAM cells: Dynamic noise margin metric. In: 2011 20th Asian Test Symposium (ATS), pp. 413–418 (November 2011)
23. Wang, J., Nalam, S., Calhoun, B.: Analyzing static and dynamic write margin for nanometer SRAMs. In: 2008 ACM/IEEE International Symposium on Low Power Electronics and Design, ISLPED, pp. 129–134 (August 2008)
24. Yang, F.-L., Huang, C.-C., et al.: 45nm node planar-SOI technology with 0.296 μm^2 6T-SRAM cell. In: Symposium on VLSI Technology. Digest of Technical Papers, pp. 8–9 (June 2004)

25. Zhang, B., Arapostathis, A., et al.: Analytical modeling of SRAM dynamic stability. In: IEEE/ACM International Conference on Computer-Aided Design (ICCAD), pp. 315–322 (November 2006)
26. Zhang, K., Bhattacharya, U., et al.: A 3-ghz 70-mb SRAM in 65-nm CMOS technology with integrated column-based dynamic power supply. IEEE Journal of Solid-State Circuits 51(1), 146–151 (2006)
27. Zhang, M., Shanbhag, N.: Soft-error-rate-analysis (SERA) methodology. IEEE Transactions on Computer-Aided Design of Integrated Circuits and Systems 25(10), 2140–2155 (2006)
28. Ziegler, J.F.: Terrestrial cosmic rays. IBM Journal of Research and Development 40, 19–39 (1996)
29. Ziegler, J.F., Lanford, W.A.: Effect of cosmic rays on computer memories. Science 206(4420), 776–788 (1979)

CMOS Implementation of Threshold Gates with Hysteresis

Farhad A. Parsan and Scott C. Smith

University of Arkansas, Fayetteville AR 72701, USA
{fparsan,smithsco}@uark.edu

Abstract. NULL Convention Logic (NCL) is one of the mainstream asynchronous logic design paradigms. NCL circuits use threshold gates with hysteresis. In this chapter, the transistor-level CMOS design of NCL gates is investigated, and various gate styles are introduced and compared to each other. In addition, a novel approach to design static NCL gates is introduced. The new approach is based on integrating each pair of pull-up and pull-down transistor networks into one composite transistor network. The new static gates are then compared to the original ones in terms of delay, area, and energy consumption. It will be shown that the new gate style is significantly faster with negligible area and energy overhead.

Keywords: NULL convention logic, NCL, C-element, threshold gate.

1 Introduction

Delay-insensitive asynchronous circuits have been the target of a renewed research effort because of the advantages they offer over traditional synchronous circuits. Minimal timing analysis, inherent robustness against power-supply, temperature, and process variations, reduced energy consumption, less noise and EMI emission, and easy design reuse are some of the benefits of these circuits [1]. NULL Convention Logic (NCL) is one of the mainstream asynchronous logic design paradigms that has been shown to be a promising method for designing delay-insensitive asynchronous circuits [2–4]. NCL circuits are correct-by-construction [3], requiring very little timing analysis, if any. In today's nanometer processes where meeting timing closure is becoming increasingly more difficult due to increasing clock rates and process variation, this quality is very attractive. NCL has been used for a number of industrial designs [4, 5], and is becoming more popular as design automation tools and techniques are being developed to automate the design process [6].

NCL circuits utilize threshold gates with hysteresis to maintain delay insensitivity. The general form of an NCL gate is very similar to a C-element [7]. Several CMOS implementation schemes have been introduced for NCL gates, including: dynamic, static, semi-static, and differential [8–10]. Each implementation offers some advantages and has some drawbacks in terms of delay, area, and power consumption. It is important for an NCL circuit designer to choose the CMOS

A. Burg et al. (Eds.): VLSI-SoC 2012, IFIP AICT 418, pp. 196–216, 2013.

Fig. 1. Symbolically complete logic concept

implementation that best fits an application. In this chapter, we introduce different CMOS implementations of NCL gates and discuss their tradeoffs. In addition, a new approach to design static NCL gates is proposed and compared with the traditional approach in terms of delay, area, and energy consumption. It will be shown that the new static gates offer faster operation, with a small increase in area, and consume almost the same amount of energy. It will be also shown that when the NCL static gates are sized for improved switching speed, the slight area disadvantage is eliminated, resulting in better speed and area.

This chapter is organized as follows: an overview of NCL logic is presented in Section 2; Section 3 discusses different CMOS implementations of NCL gates and compares them against each other; the new static gate design is then introduced in Section 4 and compared to the traditional static gate design; sizing both versions of static gates is discussed in Section 5; and these static gate styles (sized and unsized) are used to implement NCL multipliers in Section 6 to compare transistor-level simulations; and finally, Section 7 presents conclusions.

2   NCL Overview

NCL is a delay-insensitive asynchronous logic design paradigm in which control is inherent within each datum. It follows the so-called "weak conditions" of Seitz's delay-insensitive signaling scheme [11]. Similar to other delay-insensitive logic design paradigms, NCL assumes that wire forks are isochronic [12]. NCL is a "symbolically complete" logic meaning that the output validity is unambiguously determined regardless of time reference [3]. Fig. 1 shows an unknown circuit inside a black box. Assuming that the unknown circuit is a traditional Boolean combinational circuit, once the inputs are asserted, it is impossible to determine when the outputs become valid unless the circuit's delay is known. However, if the unknown circuit is using a symbolically complete logic, such as NCL, one can determine the output validity without needing to know the circuit's delay. This is because NCL uses delay-insensitive codes for data communication, alternating between set and reset phases. In the set phase, data changes from spacer (called NULL) to a proper codeword (called DATA); and in the reset phase it changes back to NULL. NCL combines DATA and NULL into a single path presented by dual-rail, quad-rail, or in general, any Mutually Exclusive Assertion Group (MEAG) signals [13].

In practice, dual-rail signal encoding is more popular, since it is most similar to traditional Boolean logic. Table 1 shows the dual-rail signal encoding. A dual-rail signal, D, consists of two wires, D^0 and D^1. D is logic 0 (DATA0) when D^0

Table 1. Dual-rail encoding

	DATA0	DATA1	NULL	Illegal
$\mathbf{D^0}$	1	0	0	1
$\mathbf{D^1}$	0	1	0	1

$= 1$ and $D^1 = 0$; it is logic 1 (DATA1) when $D^0 = 0$ and $D^1 = 1$; and it is NULL when $D^0 = 0$ and $D^1 = 0$. D^0 and D^1 are mutually exclusive, such that they are never asserted at the same time; doing so would produce an illegal codeword.

Fig. 2 shows a simple Boolean AND gate versus a dual-rail NCL circuit that performs the same AND operation. For the Boolean AND gate, inputs X, and Y, and output Z use only one wire, but the dual-rail NCL AND circuit uses two wires for each input and output. For the Boolean AND gate, initially X = 1 and Y = 0, so output Z is 0. For the NCL AND circuit, initially X is DATA1 ($X^1 = 1$, $X^0 = 0$) and Y is DATA0 ($Y^0 = 1$, $Y^1 = 0$); therefore, output Z is DATA0 ($Z^0 = 1$, $Z^1 = 0$). For the Boolean AND gate, once input Y is asserted, output Z becomes invalid until the signal propagates through the AND gate and asserts the output (in this example after 1 ns). For the NCL AND circuit, however, before input Y changes to its next DATA value, all inputs must first transition to the NULL state (i.e., all input rails must go to 0) and we must wait until the output then transitions to NULL. At this point, the circuit is ready to accept a new DATA set, so X and Y can both change from NULL to DATA1. Consequently, output Z then changes from NULL to DATA1 after some time (1 ns in this example). An NCL circuit always cycles through NULL and DATA phases so the validity of the output can always be unambiguously determined by merely looking at the output. A NULL at the output means that the output is not valid and a DATA at the output means that the output is valid. For a Boolean circuit, on the other hand, the output validity can only be determined if we know when the inputs change and the worst-case propagation delay of the circuit.

NCL circuits are comprised of 27 threshold gates with hysteresis [2]. Each gate is denoted as THmnW$w_1 w_2 \ldots w_r$ in which m is the threshold of the gate, n is the number of inputs, and w_r is the weight of input r if its weight is greater than 1. Fig. 3(a) shows the symbol of an NCL gate. For an NCL gate with no weighted inputs, the output is asserted when at least m out of n inputs are asserted. As an example, the TH23 gate asserts its output when at least two out of three inputs are asserted; therefore, assuming the inputs are A, B, and C, the set function of a TH23 gate can be expressed as F = AB + AC + BC. Fig. 3(b) shows a TH23w2 gate, where input A has a weight of two. Therefore, asserting A alone asserts the gate output. The set function of the TH23w2 gate can then be expressed as F = A + BC.

The standard NCL gate library is shown in Table 2. Since NCL gates have hysteresis, once the output is asserted, it remains asserted until all the inputs are deasserted. Hysteresis behavior is required to ensure the delay-insensitivity

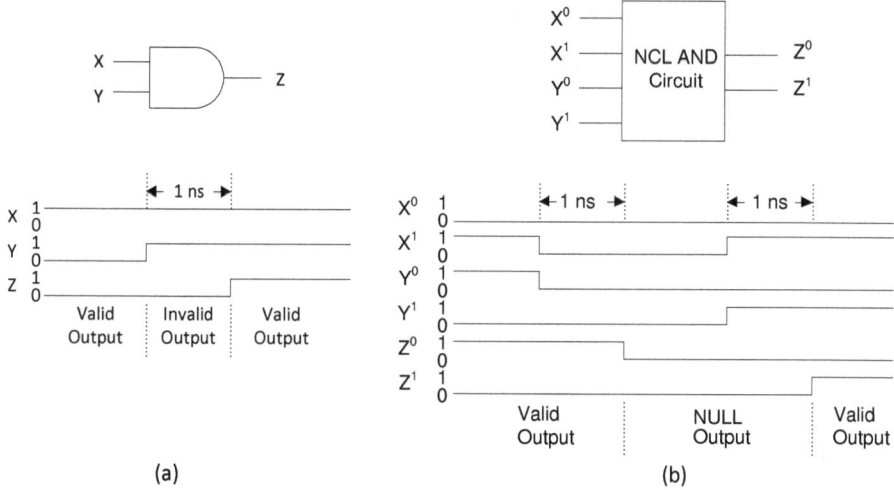

Fig. 2. (a) Boolean AND gate versus (b) NCL AND circuit

Fig. 3. (a) NCL threshold gate symbol (b) a weighted NCL threshold gate

of NCL circuits [2]. A non-weighted NCL gate with m = n (i.e., THnn) is a special case of NCL gates that is equivalent to an n-input C-element [14]. C-elements are well-known gates used in many other asynchronous logic design styles. A non-weighted NCL gate with m = 1 (i.e., TH1n) is another special case of NCL gates that is equivalent to an n-input Boolean OR gate. Among the 27 NCL gates, there are 3 gates (TH24comp, Thand0, THxor0) that are not actually threshold gates, but can be made by combining other threshold gates. These gates are included in the standard NCL gate library so that any function of 4 or fewer variables directly maps to one of these 27 NCL gates. Due to hysteresis, NCL gates act as memory elements; therefore, like any other memory element they have to be initialized. Initialization can be performed implicitly by asserting/deasserting all the gate inputs, or it can be done explicitly by adding a reset input to the gate. Depending on whether the reset signal asserts or deasserts the gate output, resettable gates are denoted with an 'n' (output deasserted) or a 'd' (output asserted) at the end of their name. Additionally, the output of an NCL gate can be provided in its inverted form; this is denoted by a small circle at the output of the gate symbol and a 'b' at the end of the gate name. Fig. 4 shows how the NCL AND circuit in Fig. 2 can be built using two NCL gates, based on the canonical SOP equations for both the rail1 and rail0 outputs,

Table 2. Standard NCL gate library

NCL Gate	Set Function
TH12	A + B
TH22	AB
TH13	A + B + C
TH23	AB + AC + BC
TH33	ABC
TH23w2	A + BC
TH33w2	AB + AC
TH14	A + B + C + D
TH24	AB + AC + AD + BC + BD + CD
TH34	ABC + ABD + ACD + BCD
TH44	ABCD
TH24w2	A + BC + BD + CD
TH34w2	AB + AC + AD + BCD
TH44w2	ABC + ABD + ACD
TH34w3	A + BCD
TH44w3	AB + AC + AD
TH24w22	A + B + CD
TH34w22	AB + AC + AD + BC + BD
TH44w22	AB + ACD + BCD
TH54w22	ABC + ABD
TH34w32	A + BC + BD
TH54w32	AB + ACD
TH44w322	AB + AC + AD + BC
TH54w322	AB + AC + BCD
THxor0	AB + CD
THand0	AB + BC + AD
TH24comp	AC + BC + AD + BD

shown in equations 1 and 2, respectively, and mapping these to the set function of the gates shown in Table 2.

$$Z^1 = X^1Y^1 \tag{1}$$

$$Z^0 = X^0Y^0 + X^0Y^1 + X^1Y^0 \tag{2}$$

Therefore, output Z becomes DATA1 when both X and Y are DATA1 and it becomes DATA0 when either input is DATA0 and the other input is DATA (i.e., DATA0 or DATA1). Reference [2] elaborates on how to design more complex combinational logic circuits using NCL.

The NCL design framework consists of delay-insensitive (DI) Combinational Logic blocks sandwiched between DI Registers. This design framework, shown in Fig. 5, is very similar to the traditional synchronous design framework, except that Completion Detection blocks are used to synchronize data communication

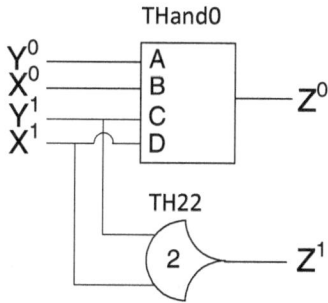

Fig. 4. NCL AND circuit

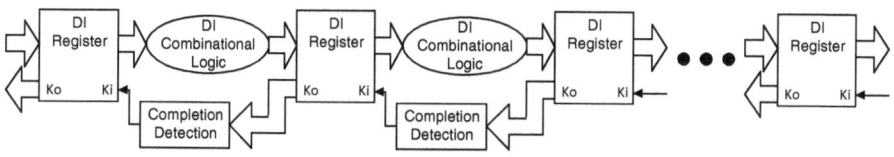

Fig. 5. NCL design framework

instead of a global clock. Completion Detection checks the output of a register to see if the previous DATA (NULL) has successfully propagated through the Combinational Logic; if so, it then allows the next NULL (DATA) to start propagating through the Combinational Logic. Ki and Ko are the handshaking signals used for requesting and acknowledging DATA and NULL. A typical DATA/NULL cycle is shown in Fig. 6. It starts with DATA propagating through a combinational block; once DATA passes the following register, the completion detection block acknowledges that DATA evaluation is finished and that NULL can now propagate. Then NULL propagates through the combinational block and clears the previous DATA; once NULL passes the register, the completion detection block acknowledges that NULL propagation is complete and allows the next DATA to start propagating through the combination block. The time period between two consequent DATA phases is called the DATA-to-DATA Cycle Time (T_{DD}), and is a measure of an NCL pipeline's throughput.

A single-bit dual-rail NCL register is shown in Fig. 7, where I^0 and I^1 are the input rails and O^0 and O^1 are the output rails. A single-bit NCL register is comprised of two TH22n gates and one inverting TH12 gate. When a combinational block is ready for DATA, Ki is asserted, allowing DATA to pass through the register; and once DATA is evaluated by the combinational block, Ki is deasserted, allowing NULL to pass through. The Ki signals of a multi-bit register are all connected together and connected to the output of the completion detection block of the next register.

The completion detection block detects whether there is a complete DATA/NULL set at the output of a register. When a register's output is NULL (i.e.,

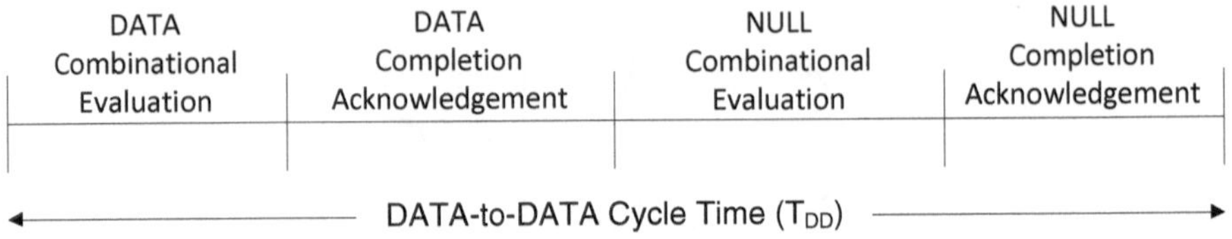

DATA Combinational Evaluation	DATA Completion Acknowledgement	NULL Combinational Evaluation	NULL Completion Acknowledgement

◄─────────────── DATA-to-DATA Cycle Time (T_{DD}) ───────────────►

Fig. 6. DATA/NULL cycle

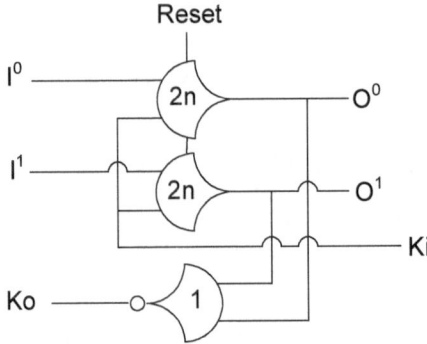

Fig. 7. A single-bit dual-rail NCL register

both output rails in Fig. 7 are deasserted), the inverting TH12 gate is asserted to request the next DATA (rfd). When a register's output is DATA (i.e., either of the output rails in Fig. 7 is asserted), the inverting TH12 gate is deasserted to request NULL (rfn). All *Ko* outputs of a multi-bit register are input to a completion detection block that asserts its output when all *Ko* signals are rfd, and deasserts its output when all *Ko* signals are rfn. An n-bit completion detection block, shown in Fig. 8, is equivalent to an n-input C-element, comprised of THnn gates. The minimum number of levels required for a completion detection block is $\lceil \log_4 n \rceil$, where n is the number of *Ko* signals [2].

3   CMOS NCL Gate Design

3.1   Dynamic Gates

The dynamic implementation of NCL gates can be used in real-time computing applications where a minimum data rate is guaranteed so that the state information can be maintained on an isolated node. The structure of an NCL dynamic gate is shown in Fig. 9(a).

The set block realizes the set function of an NCL gate, such that when the set function becomes true, the set block becomes active and discharges the internal node Y, causing output Z to be asserted. Similarly, when all inputs are deasserted, the reset block becomes active and charges the internal node Y to

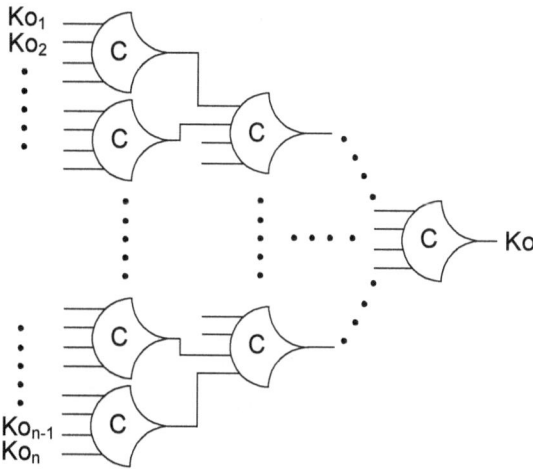

Fig. 8. NCL completion detection block

V_{DD}, causing output Z to be deasserted. In a CMOS implementation of NCL dynamic gates, the set block is a pull-down network of NMOS transistors, derived from the equations in Table 2 for each of the 27 NCL gates. On the other hand, the reset block is always a series chain of PMOS transistors consisting of one transistor per input; therefore, NCL gates that have the same number of inputs have the same reset block. The reset function of an NCL gate with n inputs can be expressed as:

$$reset = I_1' \bullet I_2' \bullet \ldots \bullet I_n' \tag{3}$$

where I_n represents input n. For most NCL gates, the set and reset functions are not complements of each other, so there are times when neither the set nor reset block is active. In a dynamic implementation, when neither is active, the internal node Y will be floating, so its value will be preserved on its parasitic capacitance, $C_{parasitic}$, for a few milliseconds before its charge leaks away, enabling the NCL gate to maintain its state, but only for a finite amount of time. Therefore, once the set function becomes true and the output is asserted, it remains asserted until the reset function becomes true and deasserts the output (hysteresis behavior). Fig. 9(b) shows the dynamic implementation of a TH23 gate, whose set function is:

$$F = AB + AC + BC \tag{4}$$

The set function can then be factored to reduce the number of transistors:

$$F = A(B + C) + BC \tag{5}$$

The NCL dynamic implementation is the smallest and fastest NCL gate style, and consumes the least amount of energy; however, since its output cannot

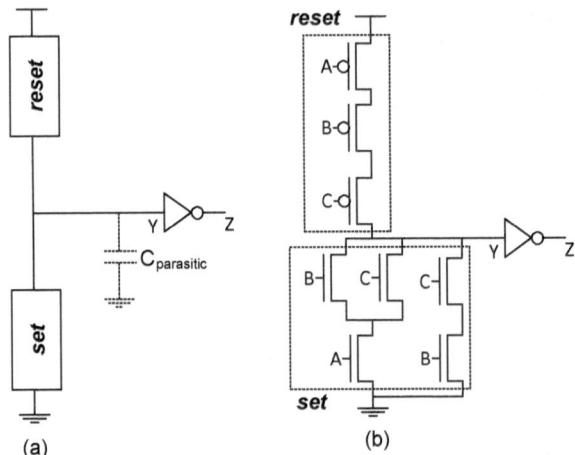

Fig. 9. (a) Structure of NCL dynamic gates (b) TH23 dynamic gate

be held indefinitely when neither set nor reset is active, it is not considered a delay-insensitive solution. Moreover, since the state information is stored on a small parasitic capacitance, it is very vulnerable to noise and charge sharing effects, although the latter can be alleviated by transistor reordering in the pull-down network [8], careful transistor sizing, and post-layout simulations. For these reasons, dynamic NCL gates are rarely used in real applications.

3.2   Semi-Static Gates

The semi-static (or pseudo-static) implementation of NCL gates utilizes feedback to maintain state information, and therefore, does not require a minimum input data rate, since it can hold the output state indefinitely. The structure of an NCL semi-static gate is shown in Fig. 10(a). In a semi-static implementation, the state information is maintained via a staticizer, in the form of a weak feedback inverter. The weak feedback inverter compensates for the leakage current that discharges the internal node Y when both set and reset blocks are inactive. This implementation is also more robust to noise and charge sharing effects because the weak feedback inverter, if carefully sized, can restore the value on the internal node Y in a reasonably short time. The semi-static implementation of a TH23 gate is shown in Fig. 10(b).

Appropriate weak feedback inverter sizing is essential for correct operation of a semi-static gate. If a feedback inverter is made very weak, it will not be able to compensate for the leakage current on the internal node, and consequently, the charge on internal node Y will leak away and the gate output Z may become invalid or switch value altogether. On the other hand, a feedback inverter that is not weak enough will require a large contention current from the pull-down network (set block) or pull-up network (reset block) to switch the output value,

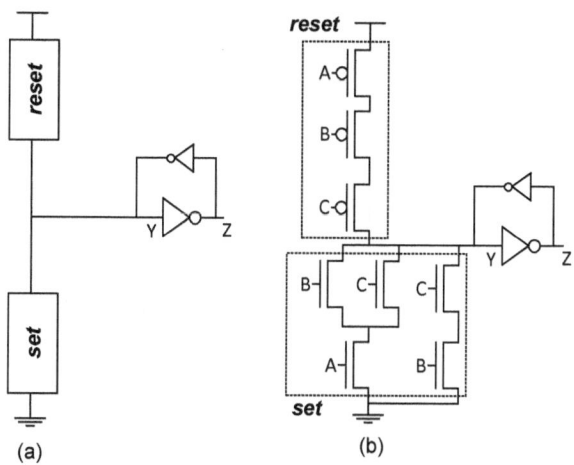

Fig. 10. (a) Structure of NCL semi-static gates (b) TH23 semi-static gate

in which case the gate's output may get stuck at a high or low value. The appropriate feedback inverter sizing also determines the performance of the gate. The weaker the feedback inverter, the more similar the semi-static implementation is to the dynamic implementation; therefore, it would be faster and would consume less energy. But, similar to the dynamic implementation, making the feedback inverter very weak makes the gate more vulnerable to noise and charge sharing effects. A more analytical discussion of semi-static C-elements, which are a special case of semi-static NCL gates, can be found in [7].

There are different ways of weakening the feedback inverter. In the standard way, shown in Fig. 11(a), usually the length of the NMOS transistor in the feedback inverter is increased. This makes the feedback inverter weak enough to be overpowered by the reset block PMOS transistor chain. The length of the PMOS transistor in the feedback inverter can also be increased or left minimum-sized since the set block pull-down network (PDN) is made of NMOS transistors and, due to the higher mobility of NMOS transistors compared to PMOS transistors, the PDN is usually able to overpower the weak inverter's PMOS transistor. Besides increasing the length of the feedback inverter's NMOS transistor, sometimes it is better to increase the width of the reset block PMOS transistor chain. The minimum set of transistors that usually need to be sized in a standard semi-static gate is shown with dashed circles in Fig. 11(a). In order to save area, sometimes it is better to add series transistors with the feedback inverter [15]. This weakening method is shown in Fig. 11(b). Here, the added series transistors limit the current available to the feedback inverter, making it weaker. The minimum set of transistors that usually need to be sized is shown with dashed circles. Finally, one can save even more area by using diode-connected transistors in series with the feedback inverter, as shown in Fig. 11(c) [16]. Using this method, the feedback inverter becomes weak enough even with minimum-sized transistors; therefore, no sizing is usually required. Again, weakening the feedback

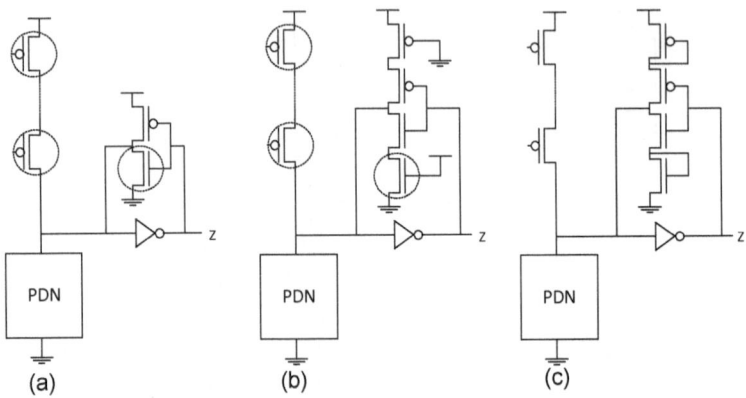

Fig. 11. Different feedback inverter weakening methods (a) standard method (b) using current limiters (c) using diode-connected current limiters

inverter makes the gate faster and less energy hungry, but the gate becomes more vulnerable to noise and charge sharing effects, so a trade-off is involved. In practice, optimal sizing of the feedback inverter is not trivial; a more analytic sizing approach is described in [17].

Among the other implementations of the NCL gates (except dynamic implementation), semi-static gates are usually considered to be small (i.e., having minimal number of transistors) and low-energy; however, this image of semi-static NCL gates significantly depends on the weak feedback inverter sizing. The relative sizing requirements for semi-static gates makes this implementation less robust to PVT variations. Also, due to the inherent contention between the set/reset blocks and the weak feedback inverter for switching the output, this implementation is usually slower than the other implementations. This contention can be minimized by appropriate weak feedback inverter sizing, but it can never be removed. A comparison of various semi-static implementations with the other implementations can be found in [16].

3.3   Differential Gates

The differential implementation of NCL gates [9] [15] is most similar to a Differential Cascode Voltage-Switch Logic (DCVSL) implementation of Boolean gates [18], with the exception of using cross-coupled inverters instead of cross-coupled PMOS transistors. A differential NCL gate is shown in Fig. 12(a). The major difference between the semi-static implementation of NCL gates and the differential implementation is that the reset block is now connecting output Z to ground through a pull-down network. Due to this change in the circuit structure, the reset block should use NMOS transistors instead of PMOS transistors, and therefore requires the input complements instead. Since each differential NCL gate provides both output Z and its complement, $\overline{Z}$, no extra logic is necessary to invert inputs. Fig. 12(b) shows the differential implementation of a TH23

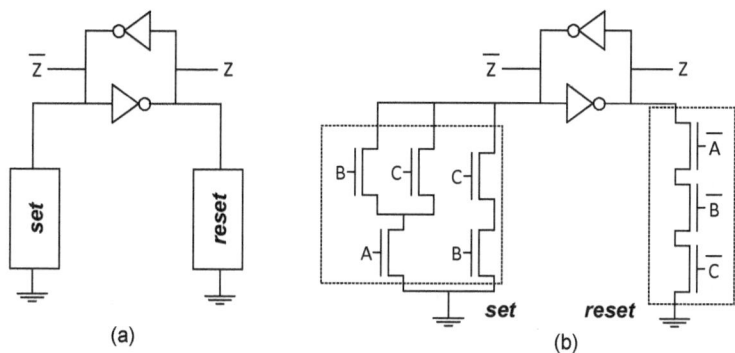

Fig. 12. (a) Structure of NCL differential gates (b) TH23 differential gate

gate. In a differential NCL gate, asserting an output requires pulling the other output low through a pull-down network (either set or reset block); therefore, before outputs switch value, there is always a short time when both outputs become low. Since in a circuit realized with differential NCL gates, the inputs of each differential gate come from the outputs of other differential gates, this ensures that before a pull-down block becomes active, the other pull-down block becomes inactive first, therefore, no contention between pull-down blocks will ever happen.

Enabling the reset block to use higher-mobility NMOS transistors instead of PMOS transistors improves the differential implementation in several ways. These improvements are mainly because of the reset block being stronger than before so it can switch the state of the cross-coupled inverters with less effort. The immediate result being that the differential implementation is usually faster than the semi-static implementation. Also, less aggressive sizing is now required, so the differential implementation is usually smaller than the semi-static implementation. In fact, a differential NCL gate can usually use all minimum-sized transistors and still function correctly. In addition, due to the symmetry of the differential implementation, the cross-coupled inverters are usually sized equally and the whole structure is therefore less sensitive to sizing, and consequently, more robust to PVT variations.

3.4   Static Gates

All the CMOS NCL gate implementations discussed so far rely on either a parasitic capacitance to maintain state information, such as in the dynamic implementation, or rely on a simple feedback mechanism via an inverter, such as in the semi-static and differential implementations. As discussed, relying on the parasitic capacitance makes NCL gates vulnerable to leakage, noise, and charge sharing problems, and eliminates their delay-insensitivity, while a feedback inverter slows down the gates due to the intrinsic switching contention involved. A static NCL gate implementation removes all these drawbacks, offering faster and more reliable operation.

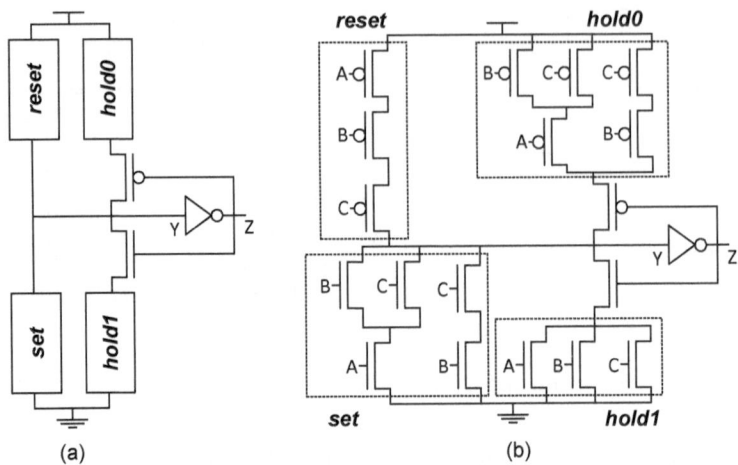

Fig. 13. (a) Structure of NCL static gates (b) TH23 static gate

As depicted in Fig. 13(a), static NCL gates are comprised of 4 transistor networks: set, reset, hold1, and hold0. Similar to other implementations, the set block determines the gate's functionality as one of the 27 NCL gates. Once the set function becomes true, the output is asserted. The output then remains asserted through the hold1 block until all inputs are deasserted. The hold1 block is simply made by ORing all inputs together; therefore, it is the same for gates having the same number of inputs. The hold1 function of a static NCL gate with n inputs can be expressed as:

$$hold1 = I_1 + I_2 + \ldots + I_n \tag{6}$$

where I_n represents input n. Since both set and hold1 blocks contribute to asserting Z and maintaining its assertion, the set equation of a static NCL gate can be described as:

$$Z = set + \left(Z^- \bullet hold1\right) \tag{7}$$

Where Z^- is the previous output value of the gate and Z is the new output value. As an example, as depicted in Fig. 13(b), the TH23 gate has the following set and hold1 functions: $set = A\left(B + C\right) + BC$; $hold1 = A + B + C$.

In order to implement a static NCL gate in CMOS technology, the complement of Z is also required. The complement of Z, denoted as Z', is realized with reset and hold0 blocks. The reset block, similar to the previous implementations, consists of all complemented inputs ANDed together. Once all inputs are deasserted, the reset block becomes active and deasserts the output. The output then stays deasserted through the hold0 block until new input values activate the set block to assert the output again. The reset equation of a static NCL gate can therefore be described as:

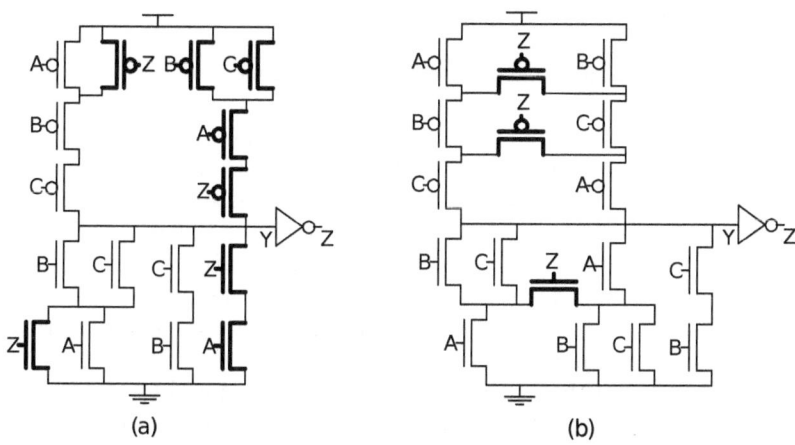

Fig. 14. (a) Original TH23 static gate (b) Proposed TH23 static gate

$$Z' = reset + \left(Z^{-'} \bullet hold0\right) \tag{8}$$

It can be proven that the following relations exist between set, reset, hold1, and hold0 functions:

$$set = hold0' \tag{9}$$

$$reset = hold1' \tag{10}$$

Equation 10 can be directly inferred from the definition of reset and hold1 functions and DeMorgan's law; and equation 9 is the logical consequence of the fact that in a static implementation, the pull-up and pull-down networks must be complements of each other to avoid a short-circuit path or a floating node. According to the above equations, the equations for a static TH23 gate are: $hold0 = A'(B' + C') + B'C'$ and $reset = A'B'C'$. The CMOS implementation of the static TH23 gate is shown in Fig. 13(b).

In contrast to the semi-static implementation, the static implementation of NCL gates is faster since output switching does not involve contention. It is also very robust to leakage, noise, and charge sharing since for any input combination the internal node Y is connected to either V_{DD} through the pull-up network, or GND through the pull-down network. Moreover, the switching threshold of static gates being typically around $V_{DD}/2$ adds to their noise immunity. Additionally, transistor sizing in a static implementation only impacts its performance, not its functionality; therefore, the static implementation is very robust to PVT variations. Its main drawback is the area overhead from adding hold0 and hold1 blocks. For example, in the case of the TH23 gate, the static implementation shown in Fig. 13(b) requires 20 transistors, while the semi-static and differential implementations only require 12 transistors. A more analytical discussion of static C-elements, that are a special case of static NCL gates, can be found in [7].

4 New Static Gates

In the previous section, area overhead was mentioned to be the main drawback of static NCL gates; however, sometimes it is possible to share transistors between each pair of pull-up (reset and hold0) or pull-down (set and hold1) networks to reduce area. For example, the direct static implementation of the TH23 gate, shown in Fig. 13(b), consists of 20 transistors; but after sharing transistors, the optimized implementation only requires 18 transistors, as shown in Fig. 14(a). There are two types of transistors in a static NCL gate: switchers, which contribute to switching the gate's output, and keepers, which only contribute to retaining the gate's state when neither set nor reset blocks are active. In Fig. 14(a) the keepers are shown in boldface.

The development of the new static NCL gates is inspired by the observation that in a traditional static NCL gate, the hold0 and hold1 transistor networks are only used for retaining the gate's state when neither set nor reset functions are true. In other words, the hold0 and hold1 transistor networks only contribute to holding the output state but not switching it. The idea behind the new static NCL gates is to integrate the set and hold1 transistor networks as well as the reset and hold0 transistor networks into a single composite transistor network such that it involves more transistors in output switching. Fig. 14(b) shows the application of this idea to the TH23 gate. The new gate structure differs from the original one in two ways. First, the reset network has been duplicated and rearranged, and then some extra PMOS transistors are added to realize the hold0 function by connecting appropriate nodes of the two PMOS transistor chains. Second, the hold1 function is realized by duplicating and flipping a portion of the set network and then connecting the middle nodes with an NMOS transistor. The new gate consists of 19 transistors, which is one transistor more than the original one; however, compared to the original gate, the number of keepers has been reduced from 8 to 3 (shown in boldface), while the number of switchers has increased from 8 to 14, resulting in faster switching compared to the original gate.

The correctness of the new gate structure can be easily proved using Boolean algebra. For the pull-up network, when $Z = 1$ both PMOS keepers are off so the function of the pull-up network can be expressed as:

$$A'B'C' + B'C'A' = A'B'C' \tag{11}$$

which is the same as the function of the reset block, and when $Z = 0$ both PMOS keepers turn on so the function of the pull-up network can be expressed as:

$$(A' + B')(B' + C')(C' + A') = A'(B' + C') + B'C' \tag{12}$$

which is the same as the function of the hold0 block. Similarly, for the pull-down network, when $Z = 0$ the NMOS keeper is off so the function of the pull-down network can be expressed as:

Table 3. Original complex static gates versus the new versions

$$(B + C) A + A (B + C) + BC = A (B + C) + BC \tag{13}$$

which is the same as the function of the set block, and when $Z = 1$ the NMOS keeper turns on so the function of the pull-down network can be expressed as:

$$(B + C + A) (A + B + C) + BC = A + B + C \tag{14}$$

which is the same as the function of the hold1 block.

The new gate structure also speeds-up output switching in one additional way. Careful investigation shows that the number of transistors in a series chain for holding the gate's state when neither set nor reset functions are true has increased. For example, the hold0 path that was originally going through Z→B→C is now going through B→Z→B→C, which is one transistor longer than the original path. Similarly, the hold1 path that was originally going through B→Z is now going through B→Z→B. Hence, the new gate structure's transistor chain length for hold0 and hold1 paths has increased by one transistor. This is equivalent to weaker hold0 and hold1 networks (i.e., the paths have higher resistance); therefore, the set and reset networks can switch the gate's output faster. This might look confusing since, as mentioned before, the set and hold0 (and similarly reset and hold1) networks are complements of each other such that they are

Table 4. Original C-elements versus the new versions

never asserted simultaneously; therefore, the set network never needs to over-power the hold0 network (or reset network overpower hold1). However, since at the time of switching there is a short moment when both pull-up and pull-down networks turn on and create a short-circuit path from V_{DD} to GND (similar to static Boolean gates), a pull-up (pull-down) network with higher resistance, and consequently less current flow, helps the pull-down (pull-up) network pull the internal node to GND (V_{DD}) with less effort, resulting in faster switching. The last interesting feature of the new static gate structure is that it is more symmetric than the original structure, resulting in closer output rise/fall times.

Converting traditional static gates to the new ones is not always easy and straightforward, especially for more complex gates. Additionally, although in the case of the TH23 gate there was only one transistor overhead for the new gate style, sometimes area overhead is more than a few transistors, resulting in an area versus delay tradeoff. Based on how complex the gate is, sometimes it is possible to partially apply this technique (e.g., to only the pull-up or pull-down network, or even just a portion of them). Table 3 shows the design of a few complex NCL gates using both the original and the new method, with keeper transistors

Table 5. Comparison between original and new static gate styles

Gate	TPLH [ps]			TPHL [ps]			Energy [fJ]			Transistors		
	New	Original	Improve	New	Original	Improve	New	Original	Improve	New	Original	Overhead
TH22	155	168	7.70%	83	123	32.20%	18.4	18.5	0.60%	12	12	0
TH33	174	197	11.40%	128	193	33.90%	20.2	19.4	-4.50%	18	16	2
TH44	198	226	12.00%	183	262	30.20%	23.1	20.2	-14.30%	26	20	6
TH44w2	200	214	6.40%	179	198	9.70%	22.3	20.6	-7.80%	25	22	3
TH23	172	180	4.50%	115	207	44.30%	20	20.3	1.40%	19	18	1
TH34w2	191	194	1.70%	150	222	32.10%	21.6	20.3	-6.30%	27	22	5
TH24comp	160	188	14.80%	134	217	38.20%	19.8	20.4	3.00%	20	18	2
THxor0	167	189	12.00%	142	255	44.20%	20.4	20.7	1.80%	23	20	3
TH22n	167	189	11.50%	86	138	37.80%	18.7	18.9	1.30%	16	16	0
THand0	180	195	7.50%	195	252	22.80%	20.6	21.2	2.80%	21	20	1
Average	177	194	9.00%	139	207	32.50%	20.5	20.1	-2.20%	20.7	18.4	2.3

shown in boldface. The first row of gates pertains to the original design, while the second row shows the new designs. Comparing the new versions with the original ones shows that the number of keepers has been reduced in all the new versions. For the THand0 gate, the pull-up network could not be converted to utilize fewer keepers, so it is not changed. Table 4 compares the original and the new static C-elements. For the TH22 gate, the new version is equivalent to the symmetric C-element design in [12]. As mentioned before, converting the original static design to the new one is not always easy and does not follow strict rules. However, the following guidelines are helpful:

1. Remove the hold1/hold0 networks from the original design
2. Duplicate the set/reset networks
3. Rearrange/flip the duplicated networks and connect their internal nodes to the original network by adding keepers such that the hold1/hold0 functionality is ensured
4. If the new structure requires more keepers in the pull-up or pull-down networks then try to apply this technique partially or just use the original design

Table 5 shows a comparison between the new gates and the original ones in terms of delay, area, and energy consumption. The gates are implemented and simulated using the IBM CMOS9SF 90nm CMOS process. All simulations are performed under the following conditions: typical process corner, nominal power supply voltage of 1.2 V, temperature of 27 °C, and capacitive load of 10 fF. Both high-to-low (TPHL) and low-to-high (TPLH) propagation delays are included in this table. The simulation results show that on average the new gates offer 9%

improvement in TPLH and 32.5% improvement in TPHL, with a 2.2% increase in energy consumption and an average of 2.3 additional transistors per gate. For the results in Table 5, all transistors are minimum-sized and the results are averaged over all possible input combinations.

5   Sizing New Static Gates

The new static gate design speeds up switching with a reasonable area overhead (2.3 transistors per gate). However, the new static gates have the potential to be smaller than the original static gates when both gate styles are properly sized for faster switching. For example, assume that the TH23 gates in Fig. 14 need to be sized. Since only the switchers are responsible for output switching, one can double their width while allowing the keepers to stay minimum-sized. This is shown in Fig. 15. The keepers are all minimum-sized (1X) in this figure, so their size is not shown. The size of the switchers in the original static gate, however, has doubled, even for the parallel switchers, in order to account for when only one of them contributes to output switching. The switchers in the new static gate are then sized such that they provide the same pull-up/pull-down resistance as the original static gate on the switching paths. Finally, the output inverter for each gate can be sized such that it offers a balanced output rise/fall time targeting a certain output load. Assuming that the output inverters would have almost similar (or comparable) sizes, the new static gate would be smaller than the original one, shown by adding up the size of transistors for each gate. In the case of the TH23 gate, the original gate size is 24X while the new gate size is 19X.

6   Simulation Results

In order to measure the performance of the new static gate style at the circuit level and compare it to the original static gate style, a delay-insensitive NCL 4×4 pipelined multiplier [19] was simulated at the transistor level using each gate style. The results, averaged over all 256 input combinations, are shown in Table 6. All simulations are performed under the following conditions: typical process corner, nominal power supply voltage of 1.2 V, and temperature of 27 °C. In order to measure the minimum power supply voltage for each variation of multiplier, V_{DD} is dropped to the point where the NCL multiplier outputs wrong data or completely stalls due to deadlock [2].

For the minimum-sized gates, the multiplier using the new gate style is 27% faster and requires 5% more area, with approximately the same energy per operation and the same low-voltage operation capability. Table 6 also shows the comparison between the multipliers utilizing sized static gates. The multiplier realized with the new sized gates is now both faster (24%) and smaller (8%) than the multiplier using the original sized gates. In addition, the energy per operation is now 10% lower, but the minimum power supply voltage has increased by 22%.

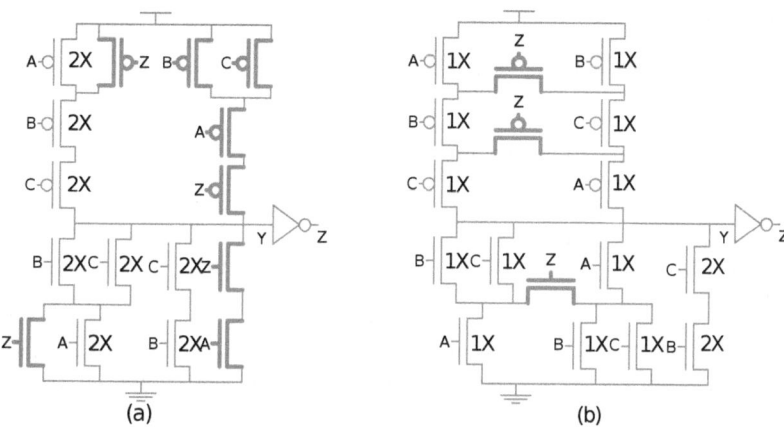

Fig. 15. A sizing example for (a) original (b) new static TH23 gates

Table 6. Comparison of NCL multipliers realized with different static gate styles

Gate Style	Original Minimum	New Minimum	Original Sized	New Sized
Delay per operation [ns]	1.45	1.05	1.29	0.98
Energy per operation [pJ]	1.29	1.28	2.62	2.34
Area [μm^2]	59.4	62.6	122.2	111.3
Minimum VDD [V]	0.25	0.26	0.22	0.27

7 Conclusion

In this chapter, different CMOS implementations of NCL gates were introduced and their trade-offs were discussed. It was shown that each implementation offers some advantages for designing NCL circuits. Omitting the dynamic implementation, since it is not delay-insensitive, comparison of the other implementations shows that static gates tend to be faster and more robust to noise and PVT variations, while semi-static gates are more energy efficient, and differential gates are more area efficient.

Additionally, a new approach to designing static NCL gates was introduced. The new gate style was compared to the original style in terms of delay, energy, and area, showing that the new gate style is significantly faster, while requiring slightly more area and energy for minimum sized gates. After sizing the gates, it was shown that the new gate style is faster, and requires less area and energy. These conclusions are supported by transistor-level simulation of a delay-insensitive NCL pipelined multiplier, to compare the different gate styles on a larger scale.

References

1. Beerel, P.A., Ozdag, R.O., Ferretti, M.: A designer's guide to asynchronous VLSI. Cambridge University Press (2010)
2. Smith, S.C., Di, J.: Designing asynchronous circuits using NULL Convention Logic (NCL). In: Synthesis Lectures on Digital Circuits and Systems, vol. 4/1. Morgan & Claypool Publishers (2009)
3. Fant, K.M.: Logically Determined Design: Clockless System Design with NULL Convention Logic. Wiley-Interscience (2005)
4. Ligthart, M., Fant, K., Smith, R., Taubin, A., Kondratyev, A.: Asynchronous design using commercial HDL synthesis tools. In: Proc. Sixth Int. Symp. on Advanced Research in Asynchronous Circ. and Syst., pp. 114–125 (April 2000)
5. McCardle, J., Chester, D.: Measuring an asynchronous processor's power and noise. In: Proc. Synopsys Users Group Conf. (SNUG). Synopsys, Mountain View, Calif., pp. 66–70 (2001)
6. Parsan, F.A., Al-Assadi, W.K., Smith, S.C.: Gate Mapping Automation for Asynchronous NULL Convention Logic Circuits. IEEE Trans. Very Large Scale Integr. (VLSI) Syst. (to be published),
http://dx.doi.org/10.1109/TVLSI.2012.2231889
7. Shams, M., Ebergen, J.C., Elmasry, M.I.: Modeling and comparing CMOS implementations of the C-element. IEEE Trans. on Very Large Scale Integ. (VLSI) Syst. 6, 563–567 (1998)
8. Sobelman, G.E., Fant, K.: CMOS circuit design of threshold gates with hysteresis. In: Proc. of the IEEE Int. Symp. on Circ. and Syst., vol. 2, 62, pp. 61–64 (June 1998)
9. Yancey, S., Smith, S.C.: A differential design for C-elements and NCL gates. In: 53rd IEEE Int. Midwest Symp. on Circ. and Syst., pp. 632–635 (August 2010)
10. Parsan, F.A., Smith, S.C.: CMOS implementation of static threshold gates with hysteresis: A new approach. In: 2012 IEEE/IFIP 20th International Conference on VLSI and System-on-Chip (VLSI-SoC), pp. 41–45 (October 2012)
11. Seitz, C.L.: System timing. In: Introduction to VLSI Systems, pp. 218–262. Addison-Wesley, MA (1980)
12. Berkel, K.V.: Beware the isochronic fork. Integr. VLSI J. 13, 103–128 (1992)
13. Verhoeff, T.: Delay-insensitive codes – an overview. Distributed Computing 3, 1–8 (1988)
14. Muller, D.E.: Asynchronous logics and application to information processing. Stanford Univ. Press, Stanford (1963)
15. Shams, M., Ebergen, J.C., Elmasry, M.I.: Optimizing CMOS implementations of the C-element. In: Proc. of IEEE Int. Conf. on Comp. Design, pp. 700–705 (October 1997)
16. Parsan, F.A., Smith, S.C.: CMOS implementation comparison of NCL gates. In: 2012 IEEE 55th International Midwest Symposium on Circuits and Systems (MWSCAS), pp. 394–397 (August 2012)
17. Li, D., Mazumder, P.: On circuit techniques to improve noise immunity of CMOS dynamic logic. IEEE Transactions on Very Large Scale Integration (VLSI) Systems 12, 910–925 (2004)
18. Heller, L., Griffin, W., Davis, J., Thoma, N.: Cascode voltage switch logic: A differential CMOS logic family. IEEE Int. Solid-State Cir. Conf. Digest of Tech. Papers, vol. XXVII, pp. 16–17 (February 1984)
19. Smith, S.C., DeMara, R.F., Yuan, J.S., Hagedorn, M., Ferguson, D.: Delay-insensitive gate-level pipelining. The VLSI Journal Integ. 30, 103–131 (2001)

Simulation and Experimental Characterization of a Unified Memory Device with Two Floating-Gates

Neil Di Spigna, Daniel Schinke, Srikant Jayanti, Veena Misra, and Paul Franzon

North Carolina State University, Department of Electrical and Computer Engineering,
Raleigh, NC USA
{nhdispig,djschink,sjayant2,vmisra,paulf}@ncsu.edu

Abstract. The operation of a novel unified memory device using two floating-gates is described through experimental characterization of a fabricated proof-of-concept device and confirmed through simulation. The dynamic, nonvolatile, and concurrent modes of the device are described in detail. Simulations show that the device compares favorably to conventional memory devices. Applications enabled by this unified memory device are discussed, highlighting the dramatic impact this device could have on next generation memory architectures.

Keywords: memory, nonvolatile, dynamic, volatile, unified, floating-gate, FLASH, DRAM, high-k dielectric, simulation.

1   Introduction

This chapter is in part based off previously published work on the demonstration of a novel double floating-gate unified memory device [1]. In this paper, that work is extended through device simulations and additional details on the fabrication, operation, and design of circuits based on such a device. Such a unified memory device could store both volatile (dynamic) and nonvolatile states simultaneously. This could have a dramatic impact on traditional memory hierarchies [2-4]. For example, the data stored in the nonvolatile mode of the device when the computer is powered down could quickly be written to the dynamic state when the power is turned on, allowing for instant-on computing. This data transfer could also operate in reverse as dynamic data could be written to nonvolatile states to allow for full or partial hibernation of the memory fabric. Alternatively, writing dynamic data quickly to nonvolatile data could enable fast in-situ checkpointing. Finally, there are a number of novel logic applications for such a device that could impact numerous areas of computation [4].

Two floating-gates (FGs) have been used previously for enhancing memory operation [5-8]. However, these designs typically have been used in an effort to increase the memory window and data retention compared to single FG devices. For example, the size of the nanocrystals in the two FG layers can be engineered to exploit the Coulomb Blockade effect [8]. In this research, however, two FGs are used to enable a device which can store both dynamic and nonvolatile states concurrently. The two modes of operation

A. Burg et al. (Eds.): VLSI-SoC 2012, IFIP AICT 418, pp. 217–233, 2013.
© IFIP International Federation for Information Processing 2013

are distinguished by the charge condition of the two FGs. The device is in the dynamic mode when charge is simply redistributed between the two FGs. It is not until charge is drawn up from the substrate that the device enters its nonvolatile mode. Therefore, the operation of the device requires the existence of a window between when charge is merely redistributed between the FGs, and when charge is drawn up from the channel; allowing for the coexistence and selective control of both states. This requires engineering the vertical stack to create and fine tune such a window. The fabrication of the proof-of-concept device is discussed in the next section and the design tradeoffs based on those decisions are considered throughout.

2   Device Fabrication and Modeling

The double FG MOSCAPs shown in Fig. 1 were fabricated to experimentally demonstrate and confirm the device operation. The process flow is outlined in Fig. 1a. The wafers were cleaned before a SiO_2 gate oxide was grown through thermal oxidation. For the bottom FG, palladium was deposited through e-beam evaporation and patterned using liftoff. HfO_2 was used as the inter-FG dielectric and deposited through Atomic Layer Deposition (ALD). The top FG was then fabricated once again using palladium. ALD was used to deposit the control dielectric of HfAlO, which has been previously shown to have low leakage that is required of ultra-scaled FLASH memory technologies [9]. A palladium control gate was deposited and patterned, followed by a backside etch and aluminum deposition. A Transmission Electron Micrograph (TEM) of the cross section of the fabricated device is shown in Fig. 1b.

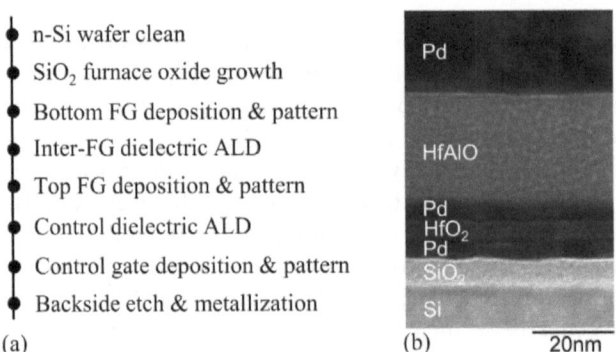

Fig. 1. Device Fabrication. (a) Recipe and (b) TEM cross section.

Being a proof-of-concept structure, the layer thicknesses of the vertical stack were not aggressively scaled, which as will be shown later in this paper have led to relatively high operating voltages. Further device scaling combined with engineering the materials has been shown through simulation to lower operating voltages, optimize device performance and achieve high storage density [4]. Towards these ends, a 65-nm gate length MOSFET, with the properties listed in Table 1, was modeled in Sentaurus TCAD.

Table 1. Device Model Properties

Layer	Material	Thickness
Control Gate	Molybdenum	10 nm
Control Dielectric	HfO_2	18 nm
Top FG	Platinum	3 nm
Inter-FG Dielectric	HfSiO	3.2 nm
Bottom FG	Magnesium	3 nm
Gate Dielectric	SiO_2	4 nm
Substrate	Bulk Si	-

This model is used to confirm the characterized device operation and is the link between the proof-of-concept structure and the circuit simulations discussed in Section 6. In addition to more aggressive thickness scaling, some important material distinctions can be made between the fabricated device and the simulated device. In contrast to the fabricated device, different materials were chosen for the two FGs, creating an asymmetry across the inter-FG dielectric, as can be seen in the energy band diagrams for the simulated and fabricated devices shown in Fig. 2. A high work function metal, Pt, was used for the top FG, whereas a low work function metal, Mg, was used for the bottom FG. This results in fast dynamic programming as electrons tunnel from the bottom FG to the top FG relatively easily. Once trapped, the deep potential well of the top FG sustains sufficiently long retention times but comes at the expense of dynamic erasing, as will be shown later in the circuit simulations. The characterization and operation of this device is discussed in the following sections, starting with the dynamic mode.

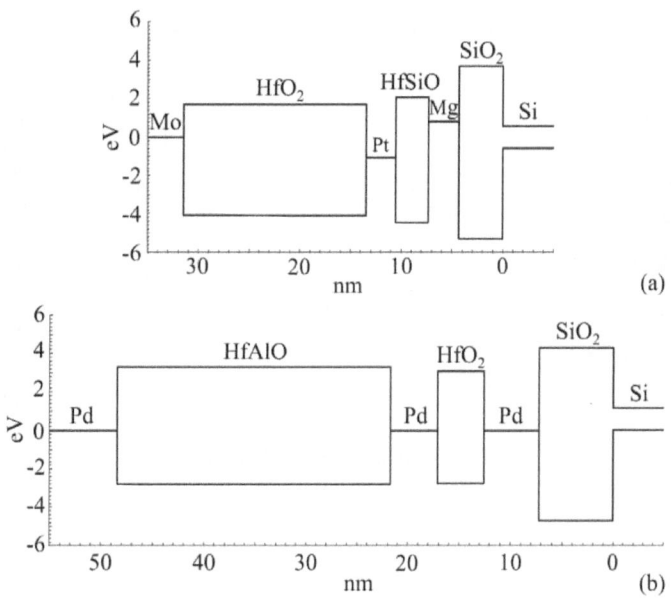

Fig. 2. The energy band diagrams of the (a) simulated and (b) fabricated devices

3   Dynamic Mode Operation

The mode of the device is determined by the applied voltage envelope. For a relatively small bias and short duration, the device's dynamic mode is programmed/erased. The dynamic operation of the device is illustrated in Fig. 3. The device is swept from a negative voltage, to a positive voltage, and then back to the negative voltage. Initially, as the device has a small negative voltage applied to the control gate, electrons will move to the bottom FG leaving behind a positive charge on the top FG, as pictured on the right-side of Fig. 3. The negative charge on the bottom FG, closer to the substrate, will cause a slight shift of the flat-band voltage to the right, as can be seen in the measured CV characteristic of the fabricated device. As the sweep continues, the voltage applied to the gate becomes positive, and the opposite charge condition results. Electrons now move to the top FG resulting in a positive charge closer to the substrate, as pictured on the left-side of Fig. 3. This positive charge closer to the substrate will cause a slight shift of the flat-band voltage to the left, once again demonstrated by the measured CV characteristic. Thus, for dynamic operation, the hysteresis is counter-clockwise, which would be the opposite direction anticipated for traditional single FG devices. Notice that for dynamic mode operation, the voltage applied to the control gate is insufficient to draw up charge from the substrate, but rather is only strong enough to simply redistribute charge on the FGs across the relatively thin inter-FG dielectric. Thus, there is no net increase in the charge on the FGs. This condition is what distinguishes the mode of operation of the device.

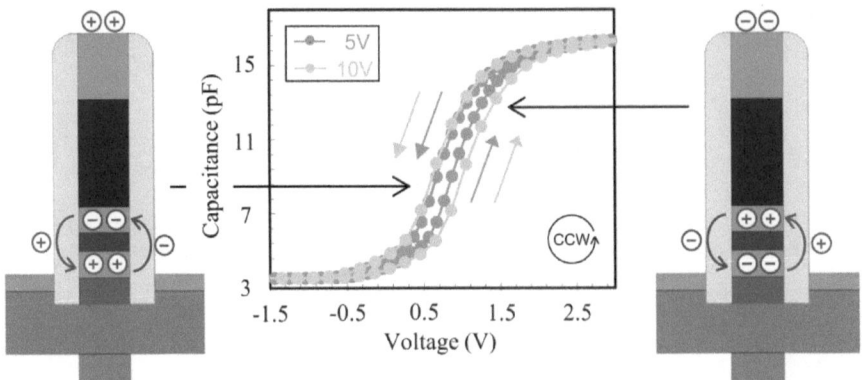

Fig. 3. Dynamic Mode Operation

The flat-band voltage shift of the fabricated device relative to the applied voltage envelope is shown in Fig. 4. As greater negative voltage envelopes are applied to the device, there is a greater positive shift in the flat-band voltage; whereas increasing positive voltage envelopes causes a greater negative shift in the flat-band voltage. The symmetry in the characteristics is indicative of the use of the same metal for the two FGs. As shown in Fig. 2b, this results in a symmetric energy barrier between the two FGs such that the program and erase characteristics are also symmetric.

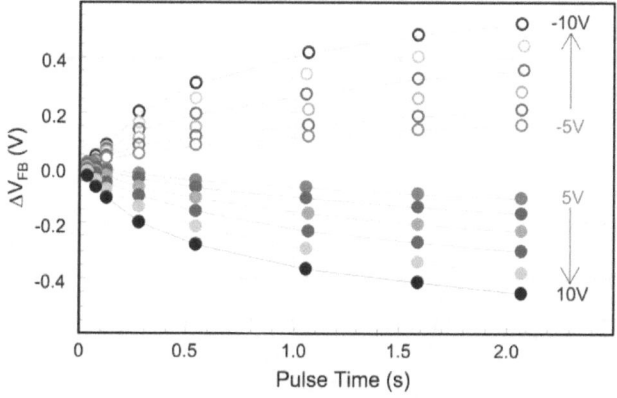

Fig. 4. Dynamic Program/Erase Characteristics

The simulations of the dynamic mode operation of the device are shown in Fig. 5. The initial uncharged device characteristics are shown (1). A 5 V pulse is applied to the control gate for 50 ns causing the threshold voltage to shift about -330 mV (2). After about 300 ms, the threshold voltage decays about 220 mV back towards the initial state of the device (3). For these simulations, it is assumed that a 100 mV difference is needed to distinguish between the two distinct states, thus a refresh is required. A 5 V refresh pulse is applied to the control gate. Since the device has not fully decayed back to the initial state, this refresh pulse only needs to be applied for 40 ns, rather than the initial 50 ns applied to redistribute charge in the fresh device. As can be seen in Fig. 5a, this refresh returns the device back to the charged state threshold voltage (4). This volatile cycle continues, requiring a refresh period of about 300 ms to retain the charged state, and thus demonstrating the dynamic mode of the device.

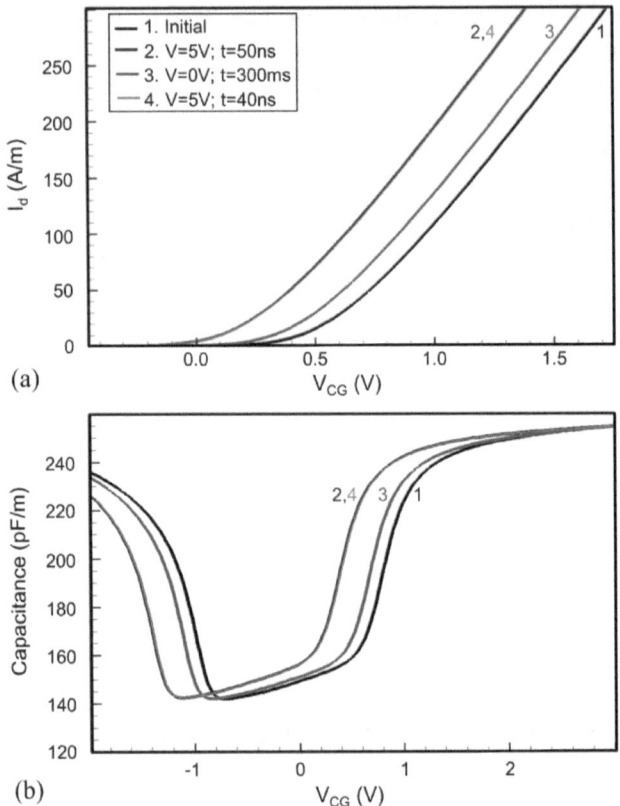

Fig. 5. Dynamic Mode Simulations. The (a) drain current and (b) capacitance vs. control gate voltage.

The volatile nature illustrated by the device simulations was confirmed in the fabricated device, as shown in the dynamic retention characteristics of Fig. 6. A +10 V pulse was applied to the fabricated device which caused a flat-band voltage shift to the left, as shown in Fig. 6a. This is directly analogous to the simulations shown in Fig. 5b. Once the bias to the control gate was removed, the CV characteristics would decay back to the original curve. This is illustrated in Fig. 6b in which the capacitance at 0.5 V is measured over time. As the charge difference between the two FGs decays, the flat-band voltage shifts to the right and the capacitance at 0.5 V decays from ~8.5 pF to ~6.4 pF. At this point, after ~22 s, the +10 V is reapplied to the control gate, once again refreshing the charge difference between the two FGs and causing the flat-band voltage to shift back to the left. This is shown for five cycles in Fig. 6b, successfully demonstrating and confirming the predicted volatile nature of the dynamic mode operation.

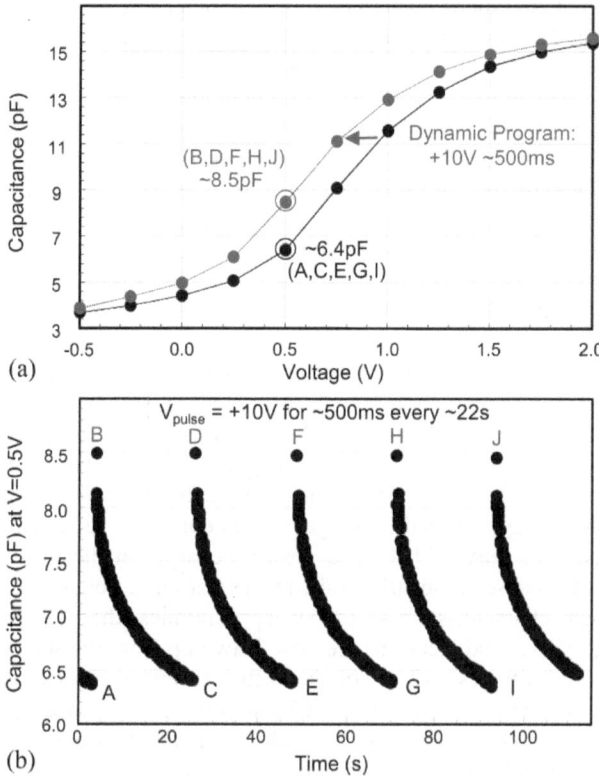

(a)

(b)

Fig. 6. Dynamic Mode Retention Characteristics. (a) A +10 V; ~500 ms pulse is applied to the control gate resulting in a negative flat-band voltage shift. (b) The capacitance was measured at 0.5 V over time for 5 cycles showing the charge difference between the two FGs decay back to the original CV curve.

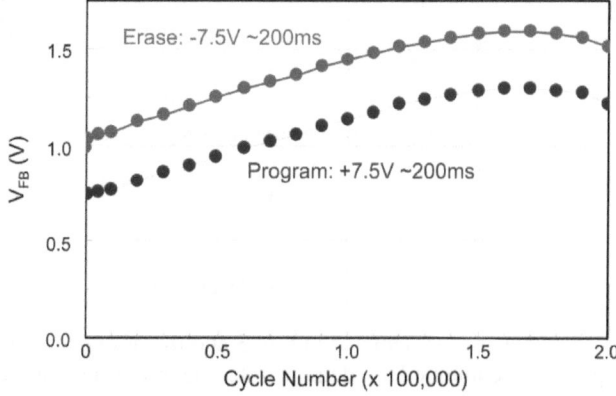

Fig. 7. Dynamic Mode Endurance Characteristics. The device characteristics are shown for 200,000 cycles with a ±7.5 V for ~200 ms pulse.

Finally, the dynamic mode endurance of the fabricated device is shown in Fig. 7. A consistent 300 mV window between the programmed and erased states is maintained as the device was cycled over 10^5 times, though a cycling drift is present. The inter-FG dielectric is critical in ensuring the stable operation over the extensive number of cycles required of DRAM. As this dielectric is further scaled, this will permit the reduction of voltages and fields, and thus the use of lower energy tunneling mechanisms that will reduce the stress on this dielectric. Choosing an appropriate dielectric for this inter-FG is actively being investigated.

4   Nonvolatile Mode Operation

The nonvolatile mode of the device is entered when a voltage pulse is applied to the control gate that is sufficient enough to draw up a net charge from the substrate to the FGs, as illustrated in Fig. 8. Once again the device is swept from a negative voltage, to a positive voltage, and then back to the negative voltage. However, unlike in the dynamic mode operation, the bias is large enough to draw up charge from the substrate. As the voltage applied to the gate starts out negative, electrons are repelled towards the substrate leaving behind a positive charge on the FGs. A net positive charge on the FGs causes a negative shift in the flat-band voltage, as can be seen in the measured CV characteristic. As the sweep continues, the voltage applied to the gate becomes positive, and electrons are now drawn up from the substrate resulting in a negative charge on the FGs. The voltage is then swept in reverse, and the negative charge on the FGs results in a positive flat-band voltage shift, once again witnessed in the measured CV characteristic. This results in a clockwise hysteresis, which is expected for traditional single FG nonvolatile devices.

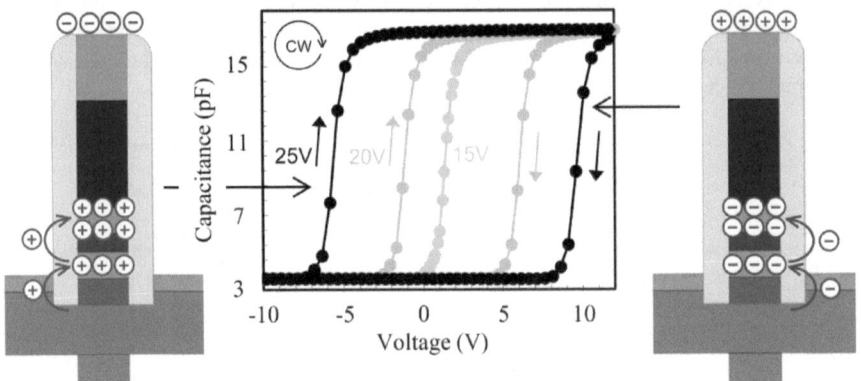

Fig. 8. Nonvolatile Mode Operation

The dynamic mode operation depicted in Fig. 3 and the nonvolatile mode operation depicted in Fig. 8 are combined to demonstrate the program/erase characteristics shown in Fig. 9. This clearly illustrates how the mode of the device is determined by the applied voltage envelope. The negative flat-band voltage window (CCW

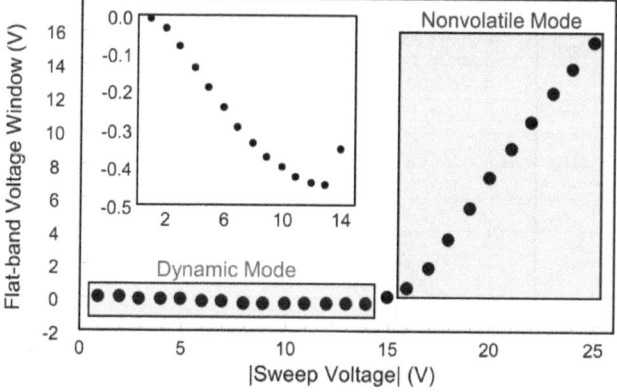

Fig. 9. Program/Erase Characteristics. The dynamic mode of the curve is enlarged in the inset.

hysteresis) is evidenced for low voltages; while the more traditional positive flat-band voltage window (CW hysteresis) occurs at higher voltages. Control devices fabricated without two FGs but with the same dielectrics did now show the dynamic mode.

The nonvolatile program/erase characteristics of the fabricated device are shown in Fig. 10. The asymmetry between the program and erase voltages is expected since, as shown in the band diagram of Fig. 2b for the fabricated device, the electrons can more easily tunnel onto the FGs then they can tunnel back to the substrate. In Fig. 10, both the initial and after-60 second flat-band voltage shifts are plotted, and in every case, the after-60 second shift is more pronounced than the initial shift. This is due to the fact that as the bias is being applied to the control gate, more of the charge is being drawn up to the top FG, relative to the bottom FG. When the external bias is removed, the charge redistributes between the two FGs resulting in an increase in the charge on the bottom FG. Since the bottom FG is closer to the substrate, this charge redistribution leads to a greater flat-band voltage shift over time as the charge settles. This is confirmed by the simulations of the nonvolatile mode of the modeled device shown in Fig. 11.

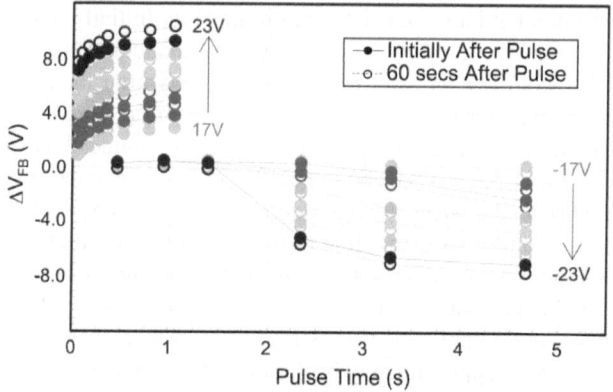

Fig. 10. Nonvolatile Program/Erase Characteristics

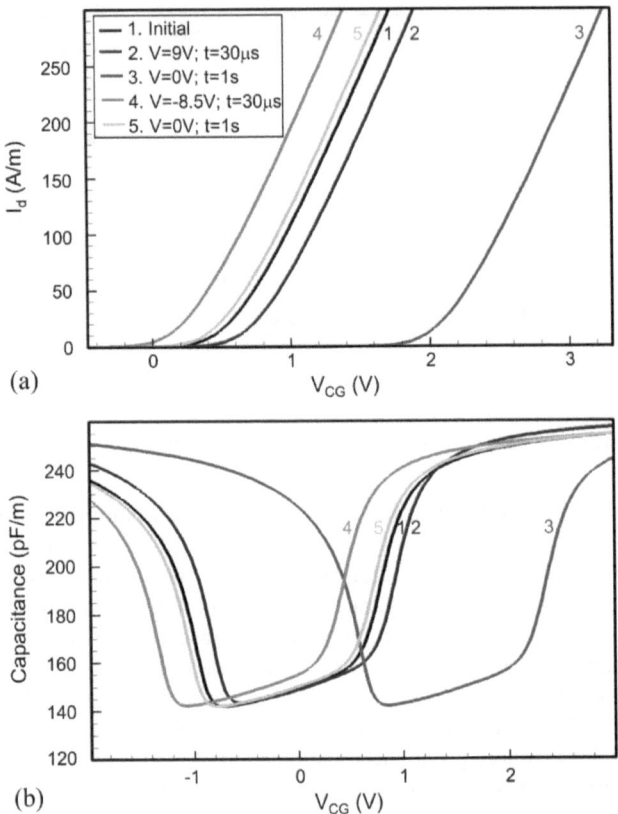

Fig. 11. Nonvolatile Mode Simulations. The (a) drain current and (b) capacitance vs. control gate voltage.

The initial uncharged device characteristics are shown (1). A 9 V pulse is applied to the control gate for 30 μs (2). This pulse is large enough to pull up charge from the channel, resulting in a net increase of charge on the FGs. Initially, most of the charge is drawn up to the top FG, limiting the impact on the channel. Thus, only a relatively minor positive threshold voltage shift occurs immediately after the pulse, as shown in Fig. 11a (2). However, after the voltage is removed from the control gate, the charge on the FGs redistributes resulting in a much more pronounced positive threshold voltage shift of about 1.52 V after about 1 s (3). This is the same phenomenon that occurred in the fabricated devices, though not to the same extent. The device does not reach its stable state until after some time passes. The relationship between the initial applied pulse and the charge redistribution settling time is currently being investigated and will have to be accounted for at the circuit level. Finally, as shown in the simulation, a -8.5 V pulse applied for 30 μs (4) followed by a charge settling period of 1 s (5) returns the device approximately back to its uncharged state.

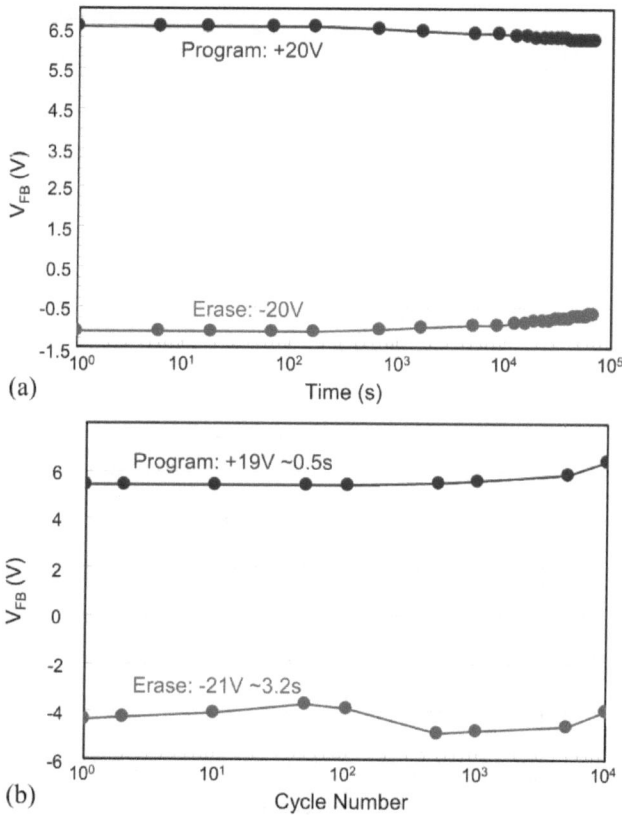

Fig. 12. The nonvolatile mode (a) retention and (b) endurance characteristics

To verify the nonvolatile nature of the fabricated device, the retention of the nonvolatile mode is plotted in Fig. 12a. A window of at least 4.5 V is maintained by extrapolating the data out to 10 years. Finally, the nonvolatile endurance of the device was demonstrated through over 10,000 cycles, as shown in Fig. 12b.

5   Concurrent Mode Operation

The device is not limited to operation in either the dynamic mode or the nonvolatile mode, but rather it can operate in both the dynamic and nonvolatile modes at the same time. The experimental verification of concurrent mode operation is shown in Fig. 13. The device is first programmed into the nonvolatile state using a +17 V sweep, as shown in Fig. 13a, which results in a positive flat-band voltage shift as negative charge is drawn up from the substrate into the FGs. To demonstrate concurrent mode operation, a dynamic state is then embedded on top of the programmed nonvolatile state by the application of a dynamic pulse of +10 V. As a response to this pulse, some of the negative charge on the bottom FG is drawn up to the top FG, leaving

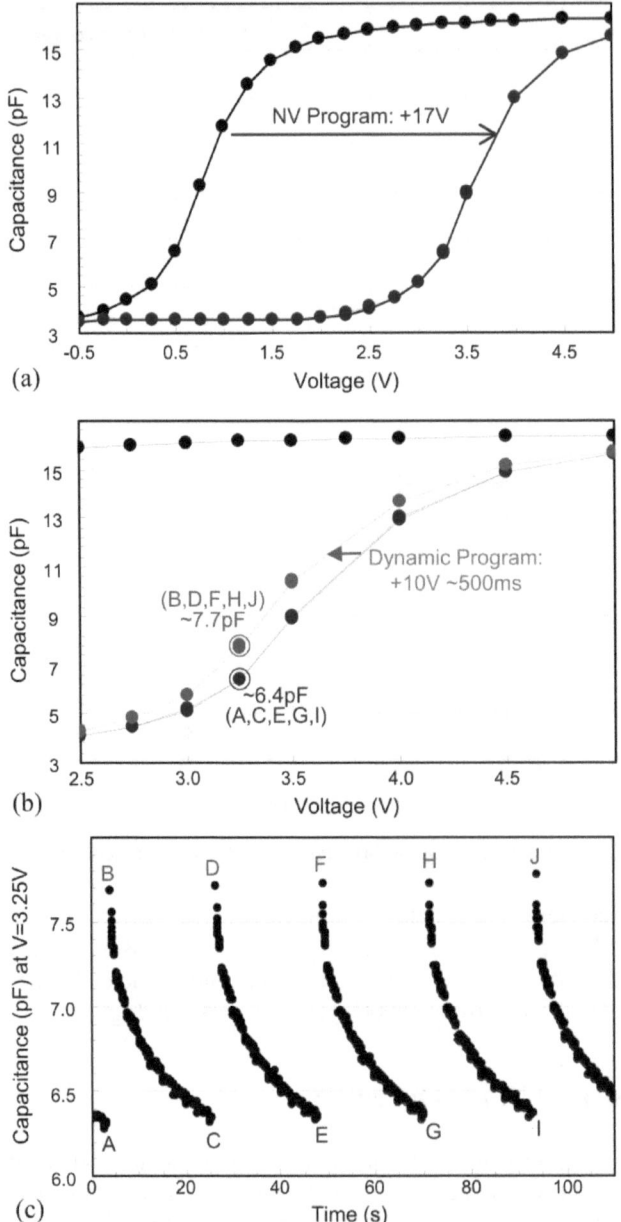

Fig. 13. Concurrent Mode Operation. (a) The device is charged into the programmed nonvolatile mode with a +17 V sweep. (b) A +10 V pulse embeds a dynamic state on top of the programmed nonvolatile state. (c) The capacitance is measured at 3.25 V over 5 cycles demonstrating the retention of the embedded dynamic state. As the dynamic state decays, the device returns to the programmed nonvolatile state, requiring a refresh.

behind a less negatively charged bottom FG, resulting in a slightly negative flat-band voltage shift relative to the original charged nonvolatile state, as shown in Fig. 13b. Once this dynamic bias is removed, the charge difference between the two FGs decays, and the flat-band voltage shifts back to the original nonvolatile programmed state. This cycling of the dynamic state embedded on top of the programmed nonvolatile state is repeated five times, as shown in Fig. 13c. Combining this data with that shown in Fig. 6, which represents the dynamic state embedded on top of the nonvolatile erased state, successfully demonstrates that the dynamic state can be embedded on both the programmed and erased nonvolatile states. Thus, concurrent mode operation of the fabricated device is experimentally verified.

The concurrent mode simulations are shown in Fig. 14. The device is first programmed into its charged nonvolatile state (1). A dynamic pulse of 5 V for 50 ns results in about a -330 mV threshold voltage shift (2). Upon cessation of the bias, the charge redistributes and the threshold voltage starts decaying back to the charged nonvolatile state (3). The CV curve of Fig. 14b is directly analogous to the experimental characterization shown in Fig. 13b. A relatively small dynamic pulse shifts the flat-band voltage in the negative direction, at which point it begins to decay back to the charged nonvolatile state flat-band voltage, as shown in Fig. 13c. Upon a refresh, the flat-band voltage once again shifts to the left (4). Thus, it is shown through simulation that a dynamic mode can be embedded on the charged nonvolatile state.

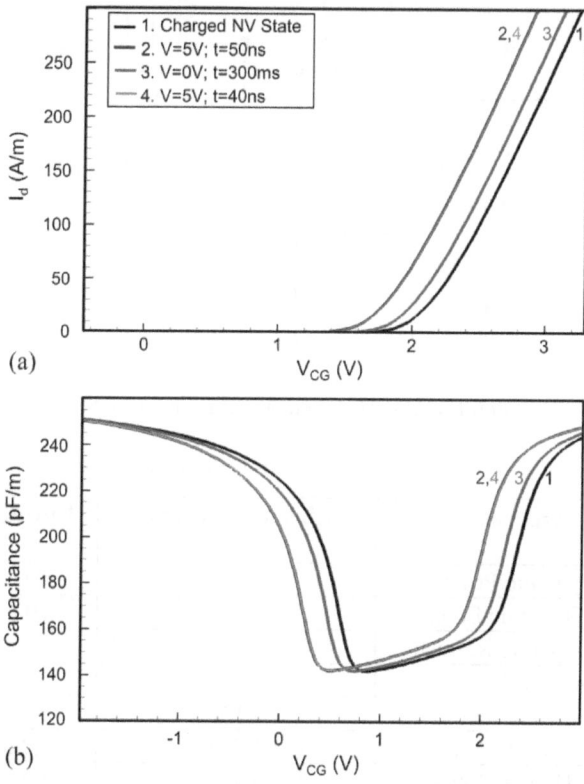

Fig. 14. Concurrent Mode Simulations. The (a) drain current and (b) capacitance vs. control gate voltage.

The dynamic, nonvolatile, and concurrent mode operation have been experimentally demonstrated. The characterization of the fabricated devices has been confirmed through device simulation, successfully verifying the operation of this novel unified memory device. A memory array using this device is discussed in the next section.

6   Circuit Simulations

The memory array shown in Fig. 15 was designed in Cadence Virtuoso 2010 using a BSIM4.0 MOSFET model with a 45-nm gate length and device parameters similar to the simulated device described in Table 1. However, instead of bulk silicon, the substrate was SOI with a thickness of 13 nm; an SiO_2 back gate dielectric of 1.2 nm was used, and the control and back gates were composed of aluminum. The operation of this memory array is described in Table 2.

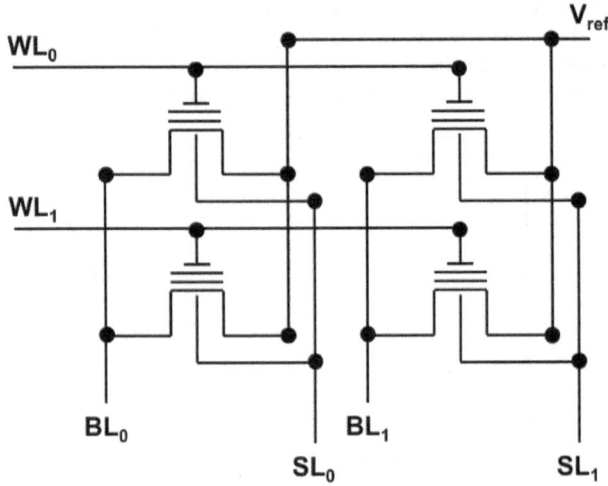

Fig. 15. Simulated memory array architecture

Table 2. Memory Array Operation

Operation	Bias Across Gate Stack	Duration
Dynamic Program	5 V	50 ns
Dynamic Erase	-5 V	10 μs
Dynamic Refresh	5 V	40 ns
Nonvolatile Program	9 V	30 μs
Nonvolatile Erase	-9 V	14 μs
Low V_t Read	1.2 V	2.2 ns
High V_t Read	2.7 V	2.2 ns

To dynamically program a target device, 3 V is placed on the appropriate WL and - 2 V is placed on the appropriate SL. This results in a 5 V bias across its gate stack, which when applied for 50 ns results in the target device being dynamically programmed, as previously described in the device simulation. However, to prevent inadvertent programming of non-target devices on that WL, the non-target SLs need to be biased to 2 V such that there is only a 1 V bias across their gate stack. This represents the dynamic retain condition.

As previously discussed, the device was engineered to have a low work function metal for the bottom FG and a high work function metal for the top FG. This resulted in an asymmetric barrier that allowed for fast dynamic programming and increased dynamic retention as charge tunneled easily from the bottom FG into the deeper energy well of the top FG, as shown in Fig. 2a. This resulted in a dynamic retention of 300 ms. However, this came at the expense of the dynamic erase; which as shown in Table 2 takes 10 μs. This is much longer than conventional DRAM. If, on the other hand, the materials are chosen to be symmetric, as was the case for the fabricated device in which palladium was used for both the top and bottom FGs, the dynamic erase time would reduce to 200 ns. Of course with a symmetric barrier, there is no longer the deeper potential well for the charge in the dynamically programmed state and so the retention time would also be reduced. However, for this device the retention time would only reduce from 300 ms to 100 ms, which could prove a wise tradeoff for reducing the dynamic erase time from 10 μs to 200 ns. Certainly further work function engineering can be performed to tailor the device performance towards target applications.

Another advantage of the device is that it operates more like an SRAM than a DRAM, and thus the read operation takes only 2.2 ns, which is much faster than DRAM. The read is also nondestructive, unlike DRAM. The memory array should also have a higher density than DRAM due to the difficulty of scaling the DRAM capacitor and maintaining sufficient charge sharing with the bitline. This device is scalable, in bulk form, to at least the 16-nm node. Through stacking, it has the potential to reach densities equivalent to the 8-nm node.

Overall, the device offers several advantages compared to conventional DRAM [3]. However, such a comparison is ill-conceived. The device may not be wholly superior to DRAM, nor to a similarly scaled single FG nonvolatile device, since it requires an extra FG and the addition of an ultra-thin inter-FG dielectric layer; but the device offers a tremendous advantage that neither of the other devices do singularly; it can store both dynamic (DRAM) and nonvolatile (FLASH) states concurrently. Such a unified memory device has enormous potential to impact next generation memory architectures.

7   Applications

There are a number of applications for such a unified memory device. For example, the device could be used to enable instant-on computing. The computer could quickly be powered down by simply moving all of the dynamic states into their nonvolatile states. If the entire memory array is written to its nonvolatile state in parallel, this would take only about 30 ms. When the user wants to power the computer back on, the memory controller simply needs to write back all of the nonvolatile data into the dynamic state. Once again, when performed in parallel, this would take only about 14 ms. In theory, the user could power up and power down the computer in only a fraction of a second. Beyond user convenience, this could allow for the operating system to power down during moments of inactivity. For example, if the user walked away from their computer to get a drink or take a phone call, the operating system could power down and conserve battery life. When the user returned, the power up penalty would only be a fraction of a second.

This device could also enable partial hibernation. For parts of the memory that are not currently being used, those arrays could be written to the nonvolatile state in the background as the user continues to operate their computer. This could enable a flexible memory fabric that could be selectively powered down which could have a significant impact on energy-proportional computing. An example application for this would be Google servers. Recently, a study on their server power usage showed that at utilization workloads that were common (20-30%), the servers operated at less than half their peak energy efficiency performance [10]. Given the nature of their utilization, current solutions to transfer to inactive modes are impractical because of both a time latency and energy penalty. The device described in this chapter could make such transitions practical by significantly reducing the wake-up penalties. Alternatively, partial hibernation enabled by this device could be used to further enhance active energy-saving schemes.

Another example application in which these devices could be beneficial is in-situ checkpointing. The device could be running continuously in dynamic mode, and then upon desire for a check-point, the entire memory array could be quickly written to the nonvolatile state in only about 30 ms. This would be much more efficient than writing through narrow channels to disk. Thus, more check-points could be efficiently taken, improving the resiliency of the computer. Upon detection of an error, the correct state could be recovered much faster than traditional memory hierarchies would permit. Instant-on computers, energy-proportional computing, and in-situ checkpointing are just a few examples of the potential that could be realized with a memory array composed of this new unified memory device.

8   Conclusion

New unified memory devices using two FGs were modeled, simulated, fabricated and characterized. The operation of the devices in dynamic, nonvolatile, and concurrent modes were demonstrated in proof-of-concept MOSCAPs and confirmed through device simulations. The programming, retention, and endurance characteristics were demonstrated for the different modes. A memory array based on these devices was designed and simulated. It was shown that these devices compare favorably to both conventional DRAM and FLASH devices. However, the true potential of these devices is not in their use as either a DRAM or FLASH replacement, but rather as a new unified memory device that can store both dynamic and nonvolatile states concurrently. Applications for such a device were discussed that highlight the significant impact this device could have on next generation memory architectures.

Acknowledgements. This article is in part based on works supported by the National Science Foundation under award nos. 0811582 and 1065458. We thank Dr. Eric Rotenberg, Steve Lipa, W. Shepherd Pitts, Shivam Priyadarshi, Vinodh Kotipalli and Narayanan Ramanan for their valuable contributions to aspects of this effort. Thanks to Dr. Dale Batchelor of AIF at NCSU for TEM analysis and to Jonathan Pierce for FIB preparation of the TEM cross sections.

References

1. Di Spigna, N., Schinke, D., Jayanti, S., Misra, V., Franzon, P.: A Novel Double Floating-Gate Device. In: IEEE/IFIP 20th International Conference on VLSI SoC, pp. 53–58 (2012)
2. Park, K.-H., Park, C.M., Kong, S.H., Lee, J.-H.: Novel Double-Gate 1T-DRAM Cell Using Nonvolatile Memory Functionality for High-Performance and Highly Scalable Embedded DRAMs. IEEE Transactions on Electron Devices 57(3), 614–619 (2010)
3. Han, J.-W., Ryu, S.-W., Kim, D.-H., Choi, Y.-K.: Polysilicon Channel TFT With Separated Double-Gate for Unified RAM (URAM)-Unified Function for Nonvolatile SONOS Flash and High-Speed Capacitorless 1T-DRAM. IEEE Transactions on Electron Devices 57(3), 601–607 (2010)
4. Schinke, D., Di Spigna, N., Shiveshwarkar, M., Franzon, P.: Computing with Novel Floating-Gate Devices. Computer 44(2), 29–36 (2011)
5. Afshari, K.: Nonvolatile Memory with Multi-Stack Nanocrystals as Floating Gates. In: 2007 NNIN REU Research Accomplishments, pp. 38–39 (2007)
6. Lee, C., Gorur-Seetharam, A., Kan, E.C.: Operational and Reliability Comparison of Discrete-Storage Nonvolatile Memories: Advantages of Single- and Double-Layer Metal Nanocrystals. In: IEDM 2003 Technical Digest, pp. 557–560 (2003)
7. Singh, P.K., Bisht, G., Hofmann, R., Singh, K., Krishna, N., Mahapatra, S.: Metal Nanocrystal Memory with Pt Single- and Dual-Layer NC With Low-Leakage Al2O3 Blocking Dielectric. IEEE Electron Device Letters 29(12), 1389–1391 (2008)
8. Ohba, R., Sugiyama, N., Uchida, K., Koga, J., Toriumi, A.: Nonvolatile Si Quantum Memory With Self-Aligned Doubly-Stacked Dots. IEEE Transactions on Electron Devices 49(8), 1392–1398 (2002)
9. Jayanti, S., Yang, X., Suri, R., Misra, V.: Ultimate Scalability of TaN Metal Floating Gate with Incorporation of High-K Blocking Dielectrics for Flash Memory Applications. In: IEDM 2010 Technical Digest, pp. 106–109 (2010)
10. Barroso, L.A., Holzle, U.: The Case for Energy-Proportional Computing. Computer 40(12), 33–37 (2007)

Author Index